TYPHOID IN UPPINGHAM:
ANALYSIS OF A VICTORIAN TOWN AND SCHOOL IN CRISIS, 1875–1877

SCIENCE AND CULTURE IN THE NINETEENTH CENTURY

Series Editor: Bernard Lightman

TITLES IN THIS SERIES

FORTHCOMING TITLES

www.pickeringchatto.com/scienceculture

TYPHOID IN UPPINGHAM: ANALYSIS OF A VICTORIAN TOWN AND SCHOOL IN CRISIS, 1875–1877

BY

NIGEL RICHARDSON

LONDON
PICKERING & CHATTO
2008

Published by Pickering & Chatto (Publishers) Limited
21 Bloomsbury Way, London WC1A 2TH

2252 Ridge Road, Brookfield, Vermont 05036-9704, USA

www.pickeringchatto.com

BRITISH LIBRARY CATALOGUING IN PUBLICATION DATA
Richardson, Nigel
Typhoid in Uppingham : analysis of a Victorian town and school in crisis, 1875–
7. – (Science and culture in the nineteenth century)
1. Public health – England – Uppingham – History – 19th century 2. Medical
policy – Great Britain – History – 19th century 3. Typhoid fever – England
– Uppingham – History – 19th century – Case studies 4. Central-local government relations – Great Britain – History – 19th century 5. Uppingham
(England) – Social conditions – 19th century
I. Title
362.1'094245'09034

ISBN-13: 9781851969913

Typeset by Pickering & Chatto (Publishers) Limited
Printed in the United Kingdom at Athenaeum Press, Gateshead

CONTENTS

Figure 1. Photograph of Edward Thring, Headmaster of Uppingham, 1853–87.
By kind permission of Uppingham School.

ACKNOWLEDGEMENTS

This study has evolved over thirty years, in three stages. Five years after I joined the staff of Uppingham School in 1971, I discovered in its library a slim volume: J. H. Skrine's *Uppingham by the Sea*, written in 1878. Given that Uppingham is in the very centre of England and therefore far from the coast, the title was eye-catching. I established that typhoid had once forced the staff of this well-known boarding school to migrate for an entire year elsewhere, taking their pupils and most of their equipment with them. I built up a narrative of events, constructed mostly from archives held within the school but also from newspaper material in Aberystwyth, mid-Wales, a coastal town 200 miles to the west of Upping-ham. This fleshed out the school's dispute with the local community leaders in Uppingham, and the extraordinary logistical and educational adventure which followed in its wake. This book is largely concerned with the first of those two themes; much has been established about the second, but a detailed study of it must await the biographer of its central character, headmaster Edward Thring.

Fifteen years later (1993) and four years after I left the school, I started to look for evidence beyond its archives, which included working through the rel-evant papers of the Local Government Board, the agency which oversaw public health matters throughout England and Wales at that time. It became apparent that more work needed to be done on the difficulties facing the leaders within the Uppingham local community. The existing picture of a heroic headmaster battling for his school against the hostility and bureaucracy of a complacent town was too simple. At that point, however, I became a headmaster myself, in Cambridge, and further research came to halt.

Following an invitation to speak to the Rutland Historical Society in April 2002, I started to read more widely, searching in the local record offices of two neighbouring county towns: Leicester and Northampton. Once I embarked on a Ph.D. thesis at University College, London (UCL) in 2004, my secondary read-ing showed how hostile public opinion in *any* community could be at that time to bureaucratic encroachments by so-called government experts. Public health officials were not exempt from this hostility.

I owe particular debt to a number of people who have assisted me along the way. First and foremost to Father Cormac Rigby, who wrote his D.Phil. thesis on Thring in the 1960s and who did much more unpublished work on him thereafter; he gave me much in time and expertise, but sadly died shortly after I completed my doctorate. Professor Anne Hardy of the Wellcome Trust Centre for the History of Medicine at UCL was a very helpful supervisor. Dr Elizabeth Hurren and Professor Alan Rogers generously advised me on the local issues, and Peter Lane helped me to piece together information about the Uppingham community in the 1870s. Jerry Rudman, the school archivist, patiently answered endless queries. Professor Bernard V. Lightman of York University, Toronto, and Mark Pollard of Pickering & Chatto in London encouraged and advised me on the preparation of this book.

The late Bryan Matthews and Geoffrey Frowde (two Uppingham colleagues) first encouraged me in this research, along with Dr Malcolm Tozer. Dr Philip Pattenden of Peterhouse, Cambridge, and Auriol Thomson of Glaston, Rutland, generously shared their knowledge of Revd Barnard Smith, and Wendy Wales advised me on her family ancestor, Revd William Wales. The search for elusive details about Dr Alfred Haviland was assisted by Professor Frank A. Barrett in Canada, Jon Edwards, Andrew Haviland and Charles Hillman in the UK, and by Chris Haviland and Brant and Virginia Taylor in the USA.

I am indebted to the staff of several record offices, especially those in Bridgwater, Leicester, Northampton, Aberystwyth and the Isle of Man. Dr John Davis in Oxford responded to my enquiry on local government financial issues, and David Sharp to another about correspondence in the *Lancet*. Phil Rayner of the Rutland Museum, Oakham, provided data about neighbouring towns to Uppingham, and Dr Paul Carter of the National Archives guided me around the Local Government Board papers.

In Cambridge I have been singularly fortunate to live close to the superb resources and staff of the University Library. Malcolm Underwood at St John's College provided further material on Haviland, and Dr Gillian Sutherland and Dr Simon Szreter gave me help early on with selecting secondary material. Dr Virginia Warren advised on medical and public health issues. Karen Gibson assisted me with typing and compilation and Julie Wilson of Pickering & Chatto at the editorial stage.

Finally I owe thanks to four Uppingham headmasters – the late Coll Macdonald (1975–82), who reduced my teaching at the start of my research in 1976; Nick Bomford (1982–91), who allowed me to pursue the investigation further; and Dr Stephen Winkley (1991–2006) and Richard Harman (2006–present) for their encouragement in the final stages. I am also indebted to the governors of the Perse School, Cambridge – who provided far more encouragement to me than Uppingham's trustees ever gave to Thring – for granting me four months' sabbatical leave in 2002–3 during which much of the material was assembled. My wife, herself one of the successors to those trustees who gave Thring so much concern, provided many constructive suggestions as the text finally took shape.

Nigel Richardson
Cambridge, April 2008

LIST OF FIGURES AND TABLES

LIST OF ABBREVIATIONS

BMJ	*British Medical Journal*
DMOP	District Medical Officer of the Poor
FRCP	Fellow of the Royal College of Physicians
FRCS	Fellow of the Royal College of Surgeons
GP	General Practitioner
JP	Justice of the Peace
LGB	Local Government Board
LRCP	Licentiate of the Royal College of Physicians
LSA	Licentiate of the Society of Apothecaries
MO	Medical Officer
MOH	Medical Officer of Health
MRCS	Member of the Royal College of Surgeons
NA	National Archives
ODNB	*The New Dictionary of National Biography*
PWLB	Public Works Loan Board
ROLLR	Record Office for Leicestershire, Leicester and Rutland
RSA	Rural Sanitary Authority
SPCK	Society for the Promotion of Christian Knowledge
UA	Uppingham (School) Archive
ULHSG	Uppingham Local History Studies Group
USA	Urban Sanitary Authority
USM	*Uppingham School Magazine*
USR	*Uppingham School Roll*

NOTES ON THE TEXT

Nomenclature

Throughout the text I have identified the Uppingham Poor Law guardians by the collective term RSA, or as the Authority. In the absence of any minute book for the sanitary sub-committee it is impossible to draw a precise distinction between its activities and those of the town's guardians as a whole. The RSA's members were responsible for their actions to central government, in the form of the Local Government Board. I have referred to this London-based body as the LGB or the Board.

Prices

Converting the prices and wages of 130 years ago accurately into modern figures is more complex than it might seem. According to the Office for National Statistics, on a composite price index with January 1974 as 100, 1875 is registered at 9.8, and 2003 at 715.2. Thus £1 in 1875 equates to roughly £75 at today's (2008) values.

English coinage at that time was divided into pounds (£), shillings (*s.*) and pence (*d.*) – 20 shillings to the pound, and 12 pence to the shilling – expressed as e.g. £12.4.8*d*.

Distances

The kilometre is approximately five-eights of a British mile. As a rough guide, distances quoted in miles should therefore be doubled in any conversion to kilometres.

WHO'S WHO IN THE NARRATIVE

Name	First Name	School Role	Town Role	Other Role
Adderley	Sir Charles	Landowner		
Bagshawe	W. A. E. V.	Housemaster		
Bell	Thomas	Medical officer	Doctor/surgeon Guardian	
Birley	T. H.	Masters' trustee		Manchester business
Brown	Frederick	Doctor/surgeon		
Brown	W. H.	Clerk to RSA Manor steward	Solicitor	
Campbell	William	Housemaster		
Candler	Howard	Housemaster		
Childs	Christopher	Sanitary officer Science master		
Christian	George	Housemaster Chaplain		
Cobb	C. W.	Housemaster		
Compton	William	Churchwarden Wine shop	Petitioner v. RSA (see Chapter 7)	
Earle	William	2nd Master Housemaster		
Evans-Freke	Hon. W. C.	Trustee	RSA chairman 1877	Landowner
Field	Rogers	Sanitary advisor to RSA		
Fludyer	Revd Sir J. H.	Trustee	Landowner	
Gainsborough	Earl of	Trustee Lord of Manor	Landowner	
Haviland	Alfred	MOH		
Hart	Wm Garner	Grocer	Petitioner v. RSA (see Chapter 7)	
Haslam	Sam	Housemaster		
Guy	J. C.	Clerk to trustees	Bank manager	Water Board secretary

Name	First Name	School Role	Town Role	Other Role
Hawthorn	John	Bookseller/printer	Water Board director	
Hodding & Beevor		Water Board solicitor		
Hodgkinson	Robert	i/c the Lower School		
Jacob	W. T.	Masters' trustee	Liverpool business	
James	Frederick	Inspector of nuisances		
Johnson	A. C.	Chairman of trustees		
Lambert	John	LGB Permanent Secretary		
Mould	John	Churchwarden		
Mullins	George	Housemaster		
Pateman	John	RSA member	Solicitor	
Rawlinson	Robert	LGB Chief engineer inspector		
Rawnsley	W. F.	Housemaster		
Rowe	Theophilus	Housemaster (to December 1875)	(later) HM Tonbridge	
Sclater Booth	George	LGB President MP		
Sheild	William	Manor steward	Solicitor	
Simon	Sir John	LGB Chief MO		
Smith	Barnard	RSA chairman	Rector, Glaston	
Tarbotton	Alfred	Engineering advisor		
Thring	Edward	Headmaster Housemaster		
Thring	Sir Henry	Parliamentary counsel		
Wales	William	Trustee	Authority member Rector	Landowner Lord of Manor
Walford	Augustus	Doctor/surgeon	Workhouse MO	
White	Charles	Ironmonger	Vote-carrier (see Chapter 7)	
Wortley	Edward	RSA member Chairman 1877	Farmer	

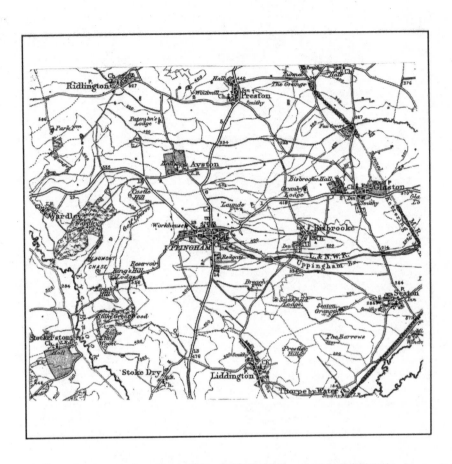

Figure 2. Villages surrounding Uppingham, reproduced from the 1-inch Old Ordnance Survey Map series, sheet 157 (Rutland and Stamford, 1906). Note that the railway did not reach Uppingham until 1894.

INTRODUCTION

The small market town of Uppingham is situated about 100 miles north of London, in the East Midlands of England. In 1875 it was unusual – although not unique – in having right at its geographical heart one of the new private, residential boarding boys' schools[1] which were springing up all over England in the wake of new-found Victorian economic prosperity.

Uppingham School consisted of just over 300 boarding pupils, aged about 13 to 18, together with a wider community of teaching and other staff and their families, who lived in a dozen or so boarding (residential) houses spread around its streets. It had an increasingly strong national reputation, and drew its pupils from newly-affluent middle-class families all over the country.

The rise of these fee-paying schools created a complex set of relationships between the schools themselves and their local communities, as the schools tended to become increasingly separated from their historical origins – as Chapter 1 will show. Uppingham's town's traders had seen what had once been their local free school become less accessible – and ultimately unaffordable – to most of them as their children reached school age. This had caused underlying tensions between the leaders of school and town which would explode when their community was struck by epidemic disease.

The typhoid outbreak of 1875–7 which ravaged Uppingham School has long been recognized as a signficant event by historians of education. Although it aroused widespread interest at the time, it has attracted little detailed historical attention since the death of its central character, headmaster Edward Thring, a decade later.

After three outbreaks of typhoid within nine months in 1875–6, and despite the school having carried out expensive improvements, Thring became convinced that the town authorities would never start a programme of major sanitary upgrading or install a mains water supply unless he forced their hand. Despite the opposition of most of the school's trustees,[2] he took a desperate gamble in March 1876 in ordering the removal of staff, pupils and much of their equipment to a temporary base 200 miles away on the Welsh coast at Borth, a straggling coastal village just north of Aberystwyth. There, in a feat of remarkable improvisation, largely funded by himself and his leading staff, he set up temporary arrangements in less than three weeks. He ran the 300-strong school in one hotel, 28 cottages and a temporary wooden assembly hall for an entire year, which included an exceptionally stormy Atlantic winter. Parents supported him to a remarkable degree; fewer than 6 pupils failed to turn up for the new term there, although most parents would never visit Borth themselves.

The epidemic has a significance which extends well beyond the school and its local area. Typhoid, and other diseases, had existed locally for many years, but only amongst the townspeople rather than in the school. However, once it began to affect the sons of the rising middle classes, it provoked a crisis of confidence amongst Uppingham School parents, who were drawn from right across the country. As a result, a dramatic chain of events was unleashed. Through these events the historian is presented with a rare case study of how a small, close-knit country community of town and school could be torn apart by politico-economic, professional and personal conflicts and rivalries at a time of crisis.

In many ways Uppingham was a typical small market town in a very traditional rural county (Rutland). It was conservative in outlook, run by tight local networks. It was an economically precarious community. In depth study of its experience in 1875–7 sheds much light on some key issues of relationships within the local community, its local government, and its local medical provision.[3] Within the town a number of powerful personalities, including Uppingham School's headmaster, were vying for leadership – especially once Thring had to face the prospect of his life's work being destroyed before his eyes. These local leaders in both town and school commanded significant powers of patronage, and their rivalries impeded sanitary progress at a time when the town authorities were already unable to satisfy rapidly rising popular expectations about public health. This stemmed in part from a lack of resources and expertise, but also from shortcomings in the relationship between government in the provinces and government at the centre – the latter symbolized by the ineffective department known as the Local Government Board (LGB).

Christopher Hamlin believes that the obstacles to radical health reform in medium-sized towns were extraordinarily complex in this period, and 'more difficult to bring about than has generally been believed'.[4] He cites reasons for this as including the need for town authorities to rely on permissive (rather than obligatory) legislation, strict treasury control from London, a lack of local technical expertise, conflicting advice from experts and a fear of making mistakes – which itself provoked the fear of legal action and writs of sequestration when errors were made. There were also inadequate enforcement mechanisms, a theme taken up by the *British Medical Journal* (*BMJ*).[5]

Much of the political and economic conflict in cities and towns centred around popular support for sanitary reform in theory, but deep resistance to raising funds to pay for it in practice – mostly via local taxes known as rates, levied on the value of properties under commercial and private ownership. Hamlin emphasizes the widespread perception amongst historians that the boards of guardians – unpaid volunteers drawn from the leading figures within local communities, and responsible for a variety of local government arrangements including public health – faced the resistance of 'a shopocracy of small businessmen, too self-interested and narrow-minded to see the long-term benefits of sewer systems and water supplies and unwilling to accept (or pay for) the expertise of engineers, chemists, or medical professionals'.[6] Other writers emphasize the narrow-minded, self-interested power and influence of landowners, the farmers' dominance on boards of guardians,[7] and the existence of a parallel and powerful middle-class elite, made up of 'medicine, veterinary science, land agency and the auctioning of property, as well as commercial services like banking and insurance'.[8]

The acute unwillingness within the members of the local governing class to pay high rates on their properties did not derive purely from economic selfishness. House-building and ownership, and the mortgage finance which backed it, could be more than a mere search for profits and rents; it was also designed to provide income for old age, for a widow or an unmarried daughter. It constituted a careful long-term savings plan.[9] Local lawyers made the arrangements and thus gained intimate knowledge of the sometimes precarious financial situation of their clients. This knowledge could make them cautious when they came to the weekly guardians' meeting and found themselves facing major spending decisions.

Yet Hamlin shows that municipal, as opposed to privately initiated, reform could be pushed through, if the guardians who made up the local urban and rural sanitary authorities (USAs/RSAs) showed determination and vision.[10] E. P. Hennock confirms this view: but it was not enough to have 'good intentions or even sound plans. Only a council which devised means wherewith to ward off a ratepayers' reaction would survive to carry out effective improvements over a long period.' This demanded careful financial administration, political skill and imagination and the possession of a substantial revenue independent of the rates.[11]

Uppingham, however, did not have men with the qualities and skills listed by Hennock, nor the other preconditions identified by Hamlin as favourable for reform: sizeable local industry, large-scale civic pride, and fear of losing out in competitive terms to rival neighbouring towns. There would be little prospect of big trading profits accruing to its guardians which could then be put towards the relief of rates. Moreover, lightly populated rural areas attracted little municipal interest in electricity, gas and tramways; the rural areas were mostly served by privately-owned businesses, and there were formidable recurrent operational costs and load repayment charges once any ambitious projects had been completed.[12]

Overall, the obstacles to improvements in Uppingham were formidable – justifying another of Hennock's assertions: that the local rating system 'could not but act as a check to any imaginative approach to the problems of urban life'.[13] This view has echoes of a much earlier one by Sidney and Beatrice Webb (some decades after 1875), who argued that there was an important role for central government in collective protection of its population in all communities: rural as well as urban.[14] They believed passionately in the benefits and virtues of municipal ownership of public services and utilities – but this belief stemmed in part from a conviction that the system of local guardians was quite inadequate to deliver those services, because of a combination of ratepayer opposition and repressive, unimaginative local leaders, inadequate to the task.[15] They were too often more oligarchic than their urban counterparts – and inexpert, too. This in turn made them afraid of placing too much reliance on their paid local officials.[16]

Most of the professional work carried out so far in the history of public health has centred around cities and large towns – notably London, Birmingham and Leeds – sometimes focusing on an individual working in an area over a long period; sometimes around civic rivalries between individual towns.[17] Even studies of the wider national picture tend to concentrate on larger communities – for example, the work of Hennock and Anthony Wohl.[18] We know comparatively remarkably little about health and mortality in the rural areas and market towns of nineteenth-century England.[19] However, while they have not generally received much attention from English historians, there were more than 400 such towns, with an average population of perhaps some 10,000 each, and two or three times that number of people in the satellite villages in surrounding areas. Their customs and attitudes shaped the minds of many millions of rural people.[20] Not only that, but such towns 'created the ethos of a French *pays*, a social and community focus which had a clear meaning for its inhabitants of "my area", as cottage industries gave way to factory production and the growth of shopping centres ...'[21]

Meanwhile there were significant local trading effects from the foundation or expansion of private boarding schools based in small towns. Many of these grew up in the wake of economic growth stemming from the Great Exhibition in London in 1851. Their development led initially to big cash and employment benefits for their local communities. T. W. Bamford shows how the market town shopocracy might be driven to overcome its low-spending instincts, if its collective livelihood were threatened by a school's hostility towards it, by a school's declining reputation or (as in Uppingham School's extreme case) a sudden misfortune threatening its very existence:

> The traders were the most vitally concerned of all local groups, and were naturally sensitive ... to anything that threatened their welfare and profits. A shop out of bounds, alterations in the boarding-house tradition, the establishment of an internal tuck-shop, bulk-buying – all these were resisted by aggrieved individuals, but never by

joint action, for that smacked too much of trade-unionism, and those who were still dependent on school trade, or hopeful of it, feared reprisal and ruin. The same fear kept the traders quiet, not only over wholesale and retail matters, but over the wider questions of health and education ... Indeed, the silence of the traders at times of crisis made up a neutral block in town affairs concerning the school, which effectively split the most worthy causes.[22]

After the 1870s – with the growth of the railways and of a more sophisticated retail network – this close economic relationship between schools and towns would begin to change: schools would rely less on local tradesmen and more on cut-price stores distributing goods from the big cities.[23] But for the time being Uppingham's townspeople faced a whole range of conflicting economic, social and political local pressures – a theme developed in detail in Chapters 1 and 2. They had to steer a delicate course between protecting their own financial situations and being loyal to their local leaders, (some of whom were their landlords), yet also being seen to support the school, which employed many of them. As in many small, enclosed (and essentially claustrophobic) communities, these pressures were exacerbated by powerful local personalities with personal prestige and territory to protect.

Popular expectations about sanitation and water supply in all areas of England were rising much faster in this period than the taxpayers' acceptance of rate increases to pay for improvements. There was so much to be done, but such inadequate governmental machinery to carry it through. In cities throughout the land, work began on radically cleansing streets following the Great Stink of London in 1858 and the pioneering work of men such as Edwin Chadwick and Joseph Bazalgette.[24] Successive governments of both parties passed radical legislation to accelerate such improvements – passing the Public Health Acts of 1848, 1872 and 1875 and setting up the Royal Sanitary Commission in 1866. Inspectorates were set up to control many areas of economic life. New medical officers of health (MOHs) such as John Simon[25] and William Henry Duncan began work, firstly in London and Liverpool, later in all large cities and finally in country areas. Meanwhile there were sweeping reforms in the registration and training of doctors. Popular expectations about better public health grew rapidly, but the means to satisfy them amounted to a sustained challenge to previous assumptions about the limits of government centralization and state interference.[26]

The rich provision of archival material about Uppingham allows us to explore all these themes in unexpected and varied detail – and to investigate the effects of new provisions for health at a significant time of legal and practical innovation. Within the span of the nineteenth century – as a whole, a time of great governmental and sanitary change – lies an interesting sub-era of less than twenty years. At its start (1871) the LGB was set up in 1871 in an attempt to ensure that public health authorities in every area carried out government's wishes.[27] Yet until the Local Government Act of 1888 set up county councils to bring a more systematic approach to local government, rural reform was far more haphazard, and local administration depended upon what Sir Edward Goschen called 'a chaos as regards authorities, a chaos as regards rates, and a worse chaos than all as regards areas'.[28]

This 'chaos' existed in the form of Justices of the Peace (JPs) responsible for local law enforcement and (by virtue of their office) also members of the boards of guardians.[29] As Justices, they too were volunteers: self-appointed and often local landowners with no directly relevant qualifications.[30] They could call on far fewer material and fiscal resources than the professional administrators who would replace them in much of local government administration after 1888,[31] and they were answerable to governments which had far fewer technological, communications and other means than nowadays to ensure that Parliament's will was carried out.

There was a central-local tension, too. Whatever London politicians might have sought to dictate from the centre, they relied on people in the localities to drive policy and its enforcement on the periphery. In retrospect, it is highly significant that it was during the two decades which spanned the first Public Health Act and the Local Government Act that Uppingham's typhoid outbreak occurred, because it illustrates the extent to which that period saw hugely increased responsibilities placed on local authorities before they were given the professional backup needed to carry them out. The difficult task of implementing central government decisions in the localities fell to the LGB, which has not enjoyed a positive reputation with historians,[32] who have emphasized its slowness and bureaucracy. However, it should not be judged purely in the light of modern expectations about the wide-ranging role of government from the centre.[33]

The dilemma over the future extent of centralization of government had been a major theme in national politics for some time, stretching right back to the Chadwick report of 1834 on the old Poor Law. Chadwick portrayed local government as ignorant, narrow-minded and corrupt, and believed that the way to progress lay through large-scale national initiative. He saw, however, that a national Poor Law organization would be expensive, and thus politically difficult to achieve – yet national legislation without local enforcement would be useless.[34] Forty years after Chadwick's report the essential conflict between effectiveness and cost had not been resolved: the LGB was under constant pressure from the Treasury, which faced hugely rising local government expenditure, and which sought to control it through requiring high interest rates to be charged on loans from the Public Works Loans Board (PWLB).[35]

As a result of this dichotomy, the LGB was faced with an impossible task. It had been 'forced to become a "Treasury" for local government, mediating the demands of the local government system on ratepayers, the money markets and the Exchequer ... an agent of boundary maintenance between local and central government ...'. It was 'obliged to control the national effects of local administrative growth, as well as to stimulate it'.[36] Elizabeth Hurren's recent work on nearby Northamptonshire[37] points to a further complication in the LGB's role: the essential confusion created in the minds of local guardians by the fact that it had two distinct departments: medical and administrative. These had combined somewhat uneasily to create an ambiguous culture which encouraged increased spending on sanitary reform on the one hand, while conducting a crusade against rising costs of outdoor relief on the other. Uppingham's experience of the LGB – as seen in the latter's dealings both with the school and the town – would reflect many of these tensions.

Uppingham's epidemic also provides us with a snapshot of the extent to which contemporary knowledge about disease in rural areas was extremely limited. The disease posed an ever-present threat to groups of people who lived in close proximity to each other in essentially closed communities such as boarding schools. They faced conflicting priorities in their everyday lives, too; for example, could the housemasters who ran the boarding houses enforce proper precautions against infection without making the lives of their boys intolerable? The dilemma about how to steer a course between what was medically sensible and what was practical day-to-day became personalized in the bitter rivalry between two doctors. On the one hand there was a traditionally-minded *laissez-faire* general practitioner, himself a former pupil of the school who must have had some sensitivity to the everyday needs of growing boys and hard-pressed staff for freedom, exercise and meeting friends; on the other, an intelligent, newly-appointed and messianic medical officer of health whose single-minded aim was to stamp out as much risk of epidemic disease as he could, with little regard for the human or economic consequences of his concern – much of it only with the benefit of hindsight – for isolation and restriction of movement.

It is hard to exaggerate the demoralizing effect which the onset of disease and the prospect of youthful death would have had on a school such as Uppingham – especially if it occurred (as typhoid usually did) at a time of year of shortening hours of daylight and increasingly wintry weather. Keeping up spirits, discipline and precautionary measures would have taxed even the ablest housemaster – who would have simultaneously been worried about the impact of disease on his livelihood. It would have produced a sharp conflict in his mind similar to that facing the school's medical officer: prudence would dictate confining pupils to houses and not letting them out to go to lessons or to visit friends. Ending term early and allowing them to go home risked spreading the infection far and wide. Yet for how long could he and his fellow housemasters enforce draconian rules on their boys?

How widespread were epidemics in the boarding schools of the time? Uppingham School suffered no major epidemic in Thring's time before 1875, although Chapter 4 will show that he had expressed a growing concern about scarlet fever and its possible causes. It had, however, experienced one recent outbreak of potentially serious disease: diphtheria, in May 1861. This was small in scale, but had resulted in the deaths of two pupils who had been nursed at school, while others who had been sent home survived.[38]

Other boarding schools had suffered far more.[39] Measures to counter disease in schools were generally reactive rather than proactive, as concentrations of young people were at risk from a variety of epidemic diseases. These included scarlet fever, diphtheria, measles, whooping cough and tuberculosis. Infections now known to be waterborne, such as cholera and typhoid, could also spread rapidly.[40] An editorial in the *Medical Officer* in 1938 referred to Rugby's experience under Arnold:

> Readers of Tom Brown's School Days will remember the description of the illness of Tom's friend, which was clearly enteric. Dr Arnold's published letters show how constantly there was present before him the spectre of sickness and death of his pupils and his great anxiety when cholera made its appearance in the Midlands.[41]

Arnold had regularly invited sixth-form reading parties to his house in the Lake District in north-west England in the holidays to refresh their health even before cholera struck, believing that the mountains and dales were 'a great point in education'.[42] He did so in the knowledge that cholera in Rugby had already caused one dispersal of his pupils back in 1832.[43]

The nineteenth century brought a major growth in boarding school numbers; living conditions were often overcrowded and spartan. Each night in the 1830s, seventy boys were locked in Long Chamber at Eton College between 8 p.m. and the following morning, as Thring himself could testify from his own schooldays there; there were no basins and no piped water; washing was done under a pump in the yard. Complaints about colds and sore throats abounded.[44] Partly as a consequence of this, Eton built a sanatorium (hospital) for its pupils in 1844. Rugby School followed suit in the early 1850s – with reason, as unhygienic piggeries, kennels and stables were part of everyday life in the town of which it formed a part. After the annual fair, the accumulated filth in the streets took over a week to remove:

> Each fetid court of beaten earth ... contained a pump for drinking water, a drain which often took the overflow from a cesspool, and a small enclosure surrounded by a low brick wall for more solid filth ... the value of the eight wide roads that radiated from the town in providing constant through winds was lessened by the ditches that ran alongside them. In these ditches the sewage from the town was collected and spread as manure upon the un-drained fields 'little better than a morass'. From ditches and cesspools the sewage of the town seeped into the drift gravels ... Held in this gravel subsoil, 'the receptacle for the chief fluid filth of the town', the water was drawn up from wells about twelve feet deep. This the inhabitants drank.[45]

Conditions inside many similar schools were little better. Westminster School's reputation had already suffered over a long period after Dean Buckland opened the drains for examination in the late 1840s;[46] cubicles were put up in the 1860s in the 'dormitory', but the rats were so numerous that they ate items of the boys' clothing as well as the food. At Winchester College in 1875 privies were still extremely basic, with their contents 'passing into a stream, called "Little Brook", which passes as a sewer in front of the college gate and receives half the town sewerage, which is abominable'.[47] Epsom College was criticized in the *Lancet* in the same year for having 'drainage until recently into large cesspits, found to be very unsatisfactory. Present system improperly ventilated. Water supply inadequate; not a constant supply.'[48]

Many years later, a committee of the Medical Research Council (1929–38) enquired into the prevalence and mode of spread of epidemics in residential schools spanning several centuries, and the growth in numbers of medical officers and sanatoria which they had recently acquired. Not surprisingly, it concluded that the illnesses had been many and varied, and it chronicled them and the temporary closures of better-known schools which had resulted from them. Typhoid was by no means the most prevalent, although Lancing was another school struck by it, in 1886.[49] The most frequent diseases included smallpox (especially before 1850) and influenza, but there were others too. Christ's Hospital suffered major bouts of ringworm in the 1830s; even sending the boys home failed to effect a cure, resulting in a vigorous campaign in the *Lancet* for better food and medical facilities. Charterhouse suffered a mumps epidemic in the 1860s. There was measles at Marlborough in 1846 and 1848 which led to the college being closed, and in addition between 1852 and 1870 it suffered twenty-six deaths, of which eight were from pneumonia, three from meningitis, four from acute rheumatism and two from appendicitis. At Radley College (a few miles south of Oxford) influenza and fever caused occasional fatalities.[50]

Haileybury, founded in 1862, built its sanatorium only four years after the college opened; and early in-patients included victims of smallpox, typhoid and scarlet fever (23 cases in 1871 and 16 in 1873).[51] In suffering from scarlet fever, it had experienced one of the two particular epidemic scourges of boarding schools. The illness was nationally prevalent in this period; the average annual death rate in England and Wales from it (per thousand persons living) rose from 0.83 in the decade 1851–60 to a peak of 0.97 in 1861–70, before falling back to 0.16 by 1900. The worst years were 1864, 1870 and 1874.[52] Creighton's *History of Epidemics in Britain* (1891–4) describes 'the enormous number of deaths [from the disease] during some 30 or 40 years in the middle of the nineteenth century [as] one of the most remarkable things in our epidemiology'.[53]

The schools could not escape scarlet fever. It was rife in Eton in the 1840s, and at Winchester, where boys had been dispersed because of it in 1843. 'Terrible illness' struck the school again a year later, and 'half the inmates were prostrated' in 1846.[54] Two sons of headmaster Moberley died there in 1858 and 1871.[55] Harrow boys were sent home twice in the 1860s after it broke out (along with other mysterious rashes which the doctors could not account for.[56] Cranleigh was hit in 1863 within a year of its opening.[57] Radley boys were sent home in 1865, and Marlborough suffered nine fatalities between 1858 and 1870[58] – the last being an epidemic which left only 150 boys at school during its peak. Wellington suffered three fatalities in the same year, and more in 1872 (as well as periodic septicaemia). Rossall, founded in 1844, suffered three deaths only six months after it opened.[59] The coming of autumn each year posed a special threat: the Rugby School doctor reported in the educational press in November 1888 that 'during the last three weeks no fewer than eleven southern schools have broken up ... owing to epidemics of scarlet fever'.[60] Even in 1896 it was a formidable disease, with a minimum of six weeks' confinement, with isolation in a room with a sheet steeped in carbolic over the door, and almost no visitors.[61]

Diphtheria was the other major threat. Again Haileybury was a victim, suffering two deaths in 1888, and many similar symptoms appeared between 1896 and 1906, resulting in seventeen boys leaving the school in one year.[62] Charterhouse suffered nine cases in 1886, after a housemaster not only ignored the advice of the school doctor to send infected pupils to the sanatorium, but also housed them on the same floor as healthy boys, and then allowed plans for a general exeat (short holiday) to go ahead.[63] Compared with these schools, Uppingham's 1861 experience of diphtheria had been comparatively minor. Fettes suffered too, and was temporarily evacuated to Windermere in the Lake District in 1883.[64] Radley suffered one fatality in 1894.[65] But diphtheria problems were at their most acute at Wellington College. Its 1883 outbreak resulted in complete drainage modernization, but this failed to prevent 'the cataclysm of 1891' – an outbreak which led to forty-one boys being admitted to the sanatorium in November, two fatalities, a crisis of confidence amongst parents and (following Thring's example in moving to Borth) temporary migration to the Imperial Hotel at Malvern, a town near the Welsh border, for a term.[66] Only with better drainage did the problem disappear.

Many preparatory boarding schools, catering for younger boys not yet old enough to go to schools such as Uppingham, had been set up in healthy areas by private owners who realized the vulnerability of boys to infectious disease. These institutions were similarly afflicted. Eagle House School suffered both scarlet fever and diphtheria in the 1850s, and Twyford suffered two deaths from the latter in 1896 before the school was evacuated to Winchester. Summer Fields in Oxford was hit by severe influenza in 1898 and the nearby Dragon School by measles in the same year.[67] The Public Schools Commission, appointed in 1864 to investigate the impact and administration of these private schools,[68] had made little reference to health, although it concluded that most schools had kept up with the domestic and sanitary advances of recent decades and that 'hardy exercise' helped to keep sickness at bay. At St Paul's School in London, however, it noted 'a great decline in the boys' health, due to overwork, fatigue, London born and bred, i.e. a delicate stock, and insufficient games and exercise'.[69]

In the Public Schools Act of 1868 there was only a single paragraph about the need for governing bodies to make regulations about the sanitation of schools. The *Lancet,* the well-known campaigning medical journal, continued its century-long fight for better conditions in such schools, forming commissions of enquiry in 1861 and again in 1875 on its own initiative.[70] These called for better hygiene and food in schools, for more comprehensive record-keeping, for parents to give notification of diseases suffered at home, and for medical examination of pupils on their return to school. The second commission, on the eve of the Uppingham typhoid outbreak, praised the new sanatoria and new water closets in some schools, but criticized poor ventilation and lighting, trapped drains and the leakage of sewer gas from town mains. It urged the appointment of medical officers in all boarding schools.

As the Uppingham crisis developed, Thring observed that although Rugby had indeed once dispersed itself into reading parties in houses all over the Lake District, no similar school had ever considered migrating en masse to a single place.[71] We do not know whether, as he said this, he was aware that the governors of Rugby, who had been considering increasing the boarding numbers at the start of Jex-Blake's headmastership in 1874, had recently commissioned the local MOH to recommend major improvements to boarding house sanitation and water supply. The housemasters at Rugby had to be persuaded to comply, but it is likely that within months events in Uppingham forced them to recognize the wisdom of the instruction.[72]

Throughout this period, some headmasters were concerned about sanitary reform, and many were intent on getting potential sources of infection removed as far from their school as possible. Could others have been more proactive? When a school's popularity waned, attacking the local authorities over faulty drainage could be a convenient diversionary tactic (as practised by Moberly at Winchester). But some were as reluctant as the local ratepayers to become involved. They ignored the problem as long as they could.[73]

Significantly, the Wellington College medical officer, Dr Barford, had complained for twenty years before the 1891 diphtheria crisis about the state of its drains, but had been dismissed by the governing body for suggesting that £20,000 was needed to put things right – following which he carried on an independent campaign in the newspapers and the *Lancet*.[74] The 1930s Medical Research Council report may have had Dr Barford in mind when it concluded that the role of the school doctor had become one of critical importance, and praised the role of the medical press in highlighting neglect or complacency.[75] Meanwhile, back in 1887, Clement Dukes, the medical officer at Rugby School[76] wrote:

> I have seen cesspools at one of the most popular and expensive schools in the kingdom in such a state of repletion that it would be impossible for the boys to use them without defiling themselves with the decomposing ordure. I may add that I saw this condition, on the occasion I refer to, on the last day of the vacation, and the state of things had existed probably since the end of the previous term.[77]

Despite the fact that the ideas of contagion and (to a lesser extent) infection circulated widely among the parent clientele of such schools,[78] large numbers of them were prepared to accept the increased risks of sending their sons away to school rather than keeping them at home, despite the knowledge that the schools had limited facilities, if any, for dealing with serious and epidemic illness. Indeed, in the middle of the century, the reaction of headmasters tended to be to send sick boys home.[79] Glenalmond's headmaster stated in 1858: 'I will *not* have boys die here'.[80]

What of our knowledge of typhoid itself? Since that date, our medical knowledge has increased a great deal. Unlike the Victorian headmasters, who knew little beyond the fact that it was acute and highly infectious, we now know that it is a systemic infection caused by the bacterium *salmonella typhi*. Untreated, it lasts between three and four weeks, killing about 10 per cent of its victims and leaving 2 per cent as permanent carriers. It is progressive – marked by the gradual onset of a sustained fever with headaches, cough, severe digestive discomfort and generalized weakness. It can also cause spleen and liver enlargement, and is sometimes marked by a rose-spot rash. The attack rate of the disease is in proportion to the number of organisms ingested. Almost unique among the *salmonellae*, its bacilli are adapted only to humans. It is normally waterborne – i.e. contracted through drinking water contaminated with the bacterium *salmonella typhi*. It is often transmitted via sewage-contaminated water, or by flies which carry the bacterium from infected faeces to food. The bacillus can survive for many weeks both in water and in ice. Rivers, ponds and wells are all infected by carriers, either directly or via excreta washed down by rains or faulty sanitary systems. Thus control depends on separating sewage and drinking water.[81] It can also be spread through contaminated food (especially by carriers handling milk, ice cream, fruit and salads, or as a result of shellfish growing in contaminated water), infected vomit and typhoid pus.[82]

The typhoid patient usually ceases to excrete the bacillus within a month of contracting the illness, but convalescent carriers may carry on doing so for up to about six months, and it can remain in chronic carriers for some years. There are also symptomless carriers – especially dangerous because their existence is often picked up only during the investigation of an epidemic, if at all. Some 3 per cent of persons who have been infected continue to excrete bacteria in either urine and/or faeces once restored to health, and thus become 'healthy carriers' who may infect others through, for example, handling foods, if hygienic precautions are not scrupulously followed.

However, 130 years ago knowledge of the disease and its causes was much more uncertain. Nineteenth-century civil servants and doctors had a broad understanding of its water-borne and milk-borne nature, but little insight into precisely how this occurred.[83] One leading nineteenth-century MOH declared in 1889: 'If there is one fact more certain than another in sanitary science, it is that enteric fever occurs chiefly and almost solely when there is an excrement-sodden condition of the soil'.[84] The nineteenth century saw the gradual rise of the germ theory (water-borne 'poison') against the miasma theory (foul air or gases) and theories of contagion (person-to-person touch). Achievements came piecemeal: Gerhard in the USA and Jenner in London published their descriptions of the different features of typhoid and typhus in the 1830s and 1840s. A decade later William Budd noted the connection between typhoid outbreaks and faecally contaminated food and water – confirmed by John Snow's medical mapping of the Broad Street pump's effects during the 1864–5 London cholera outbreak. However, it was only in the years just after the Uppingham epidemic that the key discoveries in bacteriology began to emerge. Eberth and Klebs identified the cholera and typhoid bacilli in 1880; Gaffky succeeded in culturing it four years later, and in the 1890s H. E. Durham and others devised the Widal test to diagnose it. By 1900–2 the first vaccines were available, and Robert Koch had pointed out the significance of the healthy carrier. At the 1867–9 hearings of the Royal Commission on water supply, germ theories had still been speculative. It was not clear why faecally polluted water only occasionally produced epidemic disease.[85] Even though the germ theory gathered momentum in the years that followed, there was continuing disagreement about its precise nature, and a reluctance to abandon the miasma theory altogether. Moreover, medical knowledge gained in London and elsewhere filtered down only slowly to rural areas. This explains why throughout the Uppingham epidemic several theories about its cause were pursued simultaneously.

In the 1870s, in cases of water-borne typhoid (as opposed to outbreaks caused by contaminated milk or other food) a few epidemics were dramatic, with a succession of patients rapidly affected when a normally safe water supply became seriously contaminated. Mostly, however, there was a slow, ongoing series of single cases or small groups appearing over quite a period of time, resulting from low-level pollution.[86] All but the chronic carriers were hard to identify and isolate, although in an age when nearly all domestic work and cooking was done by females, it was recognized that chronic carriers typically tended to be middle-aged or elderly women.[87] Methods of treatment were at best haphazard: depletion of blood and low diet, pouring cold water over the surface of the body, 'shaving the scalp and applying cold embrocations', or ordering that all the windows be kept open. There were herbal treatments based on hellebore root and alcohol (especially champagne) for the wealthy; elm or holly bark concoctions for the less so.[88]

In 1876 the *BMJ* estimated that about 100,000 people contracted typhoid each year – with perhaps another 40,000 undiagnosed cases. Because the average case lasted up to five weeks, it estimated that nearly 14,000 were ill at any one time, costing the country over £1m per year.[89] Estimates of deaths per year varied; the *Times* suggested 10,000–12,000, although one contemporary study of waterborne typhoid put the figure at under 9,000.[90] Optimists noted that the disease was in numerical decline. Fatalities had markedly reduced from the 21,000 of 1866, and the *BMJ* of May 1876 declared that deaths from fever in that year – 7,500 – were at their lowest annual total since 1837.[91] But the rate of decline then slowed, and the threat remained real. One MOH wrote that, despite skilled nursing and careful medical treatment, typhoid's course remained 'prolonged and perilous ... excepting diphtheria it has probably the highest death-rate of all the infectious diseases prevalent in this realm'.[92]

Typhoid was no respecter of classes or persons. Whereas louse-borne typhus, and to a lesser extent water-borne cholera, tended mostly to affect poorer city dwellers, typhoid was less confined to urban areas and could affect the highest in the land. It had claimed the life of Queen Victoria's husband Albert, the Prince Consort, in 1861 and nearly carried off her eldest son, the Prince of Wales, a decade later, when he contracted the disease whilst staying at a country house in Yorkshire.[93] There was a major outbreak amongst undergraduates at Gonville and Caius College, Cambridge, in 1873, and another claimed the lives of three Oxford undergraduates in January 1875.[94]

The extent to which knowledge about typhoid was only partial can be gauged from a leading article in *The Times* on 13 January 1876, in which the paper declined to choose between the miasma and germ theories. After describing how it attacked the intestines, it described it as 'a sort of smallpox, which affects the bowels instead of the skin ...'. In the absence of bacteriological knowledge it then stated:

> It is spread abroad chiefly by discharges from the ... intestine. These, in the natural course of things, find their way into cesspools and sewers and when they do so, they render poisonous the solid and liquid contents of these receptacles and also the gas which is evolved from them. The fever is reproduced mainly in three ways – first, by poisoned sewage obtaining direct access to drinking water, by leakage or soaking, and so being swallowed; secondly by the poisoned gas escaping from the sewers into water mains or cisterns, so that it is absorbed or dissolved by the water, and so swallowed; and thirdly by the poisoned gas making its way, through badly-trapped drains or other channels, into dwelling or sleeping rooms, and so being breathed by the occupants ...

This was an age which associated odours very closely with disease.[95] William Budd declared in 1873 that 'the poison by which this fever spreads is almost entirely contained in the discharges from the bowels', but he listed as a subsequent source of infection 'the air of the sick-room', followed by 'the bed and body linen of the patient' before he came to 'the privy and the cesspool or the drains proceeding from them'.[96] Even a medical expert as famous as Sir John Simon (the first MOH for the City of London) had once believed that typhoid was spread by 'sewer atmosphere', although shortly before 1875 he had come to accept that a more likely cause was 'molecules of excrement' and 'microscopical forms', as the new germ theory gained acceptance.[97] The *Lancet* seems to have been similarly uncertain. It reported several cases among men exposed to sewer gas,[98] in one of a dozen or so editorials and papers on typhoid which it printed during a sixth-month period in 1875.

Perhaps the most revealing glimpse of the still uncertain contemporary state of knowledge can be found in the first edition of Dukes's book, a decade later. While he stated that 'in this school [Rugby], "filth diseases" such as diphtheria and typhoid, depending mostly upon unsanitary conditions of life, such as impure water have been all but exterminated', he raised 'a wider question ... whether all infectious diseases are the result of bacteria'.[99] He also cited three views on the origins of infection or contagion current at the time: first, particles of animal origin, born and growing in the body; second, particles of fungoid nature, growing in the body but induced from without; and third,

> Particles of contagia [which] are of the nature of the Schizomyceles – i.e. the members of the lowest stratum at present known in the animate world. They are commonly called bacteria, bacilli, microzymes, vibrios, spirilla, monads ...[100]

This uncertainty about precise diagnosis cause and remedy underlay the bitter animosity which developed between GP and MOH in Uppingham. Long-standing GPs were often respected (even loved) in their local communities[101] – in contrast to the newly-appointed MOHs in rural areas, who were often seen as interlopers and busybodies in the 1870s.[102] In Dr Thomas Bell, Uppingham School possessed a medical officer who was, in many ways, an archetypal rural GP in the immediate era after national qualifications were standardized.[103] But he faced pressures of his own: Chapter 3 will show that he found himself in a fierce medical market, in a town probably over-provided with GPs.[104] He was driven on by a mixture of status and self-interest, an awareness of the threat to his economic livelihood and personal animosities towards his opponents. His determined support for the school and his inveterate letter-writing and lobbying of local and national authorities once the epidemic took hold played a significant part in influencing its outcome.

Bell may also have been motivated by a desire to make up for his initial errors and inactivity. His actions – and inactions – teach us much about the state of medical and epidemiological knowledge amongst GPs in this period on the ground, as opposed to in the medical schools and consulting rooms of the big cities.[105] He can be accused of carrying out too little proactive management of the initial crisis, and of having too little curiosity about its possible causes: of adopting a purely defensive posture in the face of parental and other criticism. But he was not alone: neither the town, school nor public health authorities addressing Uppingham's problems engaged in much speculation about the epidemiological questions. Their approach to their problems was (like many of their day) essentially pragmatic, un-dogmatic and intellectually un-enquiring.[106]

Dr Bell's key opponent in Uppingham, Dr Alfred Haviland, was a passionate MOH in his war against disease – about as far from the over-promoted former inspectors of nuisances who gained such posts in many rural areas at this time as it was possible to be.[107] It is likely that he was much influenced by past events in his life, and some accused him of overreacting against the school. It could be argued that he was not especially knowledgeable about typhoid, and that he used bombast and invective to cover up for his deficiencies, both in knowledge and procedure. Why he was so splenetic is not entirely clear, but he had probably come to view Bell's more relaxed demeanour as contemptible and negligent. He may have been frustrated too by the obstructiveness which he believed he found in so many quarters – and by impatience with the overwork of covering a huge and unwieldy rural territory spanning parts of four counties. In addition, he personified the role of the MOH as interferer in the private affairs of the rural establishment, in the minds of those who crossed him – from whose ranks Thring became his principal antagonist.

There are also questions of sanitary comparability about Uppingham. How well advanced was sanitary reform in the East Midlands of England as an area? As a town, was Uppingham lagging behind its immediate neighbours? What does its experience tell us about contemporary assumptions of the best organization to provide water to a local community?

The dominance of key individuals and their personal rivalries and animosities in a local community could be a significant factor in polarizing attitudes and making both compromise and reform programmes hard to achieve. Uppingham's local leaders, two civic-minded and well-meaning clergymen, neither of whom lacked ability, were no match for Thring's dynamism (born of desperation), extremes of mood and nervous energy. The town was too small for its future to be determined by all three of these men, all forceful in their way – with a fourth, if one includes Dr Haviland.

The Uppingham epidemic also sheds important light on the career and organizational abilities of Thring himself. In the eyes of many students of nineteenth-century education in England, he has long played second fiddle to a figure from the preceding generation of school builders: the famous Dr Thomas Arnold, headmaster of Rugby School from 1828 to 1842. Thanks to the high-minded portrayal of him in Thomas Hughes's famous novel *Tom Brown's Schooldays*, published in 1857, and a string of highly laudatory biographies in the century after his death, Arnold is widely remembered as the key figure in the development of the Victorian public schools and the founder of the modern system: a reformer noted for adding mathematics, modern languages and modern history to the traditional classical curriculum,[108] and his monitorial (prefect) system which was adopted by most English secondary schools. A number of Arnold's staff went on to be headmasters elsewhere, spreading his fame and influence as they went.

By contrast, the greatest headmaster of the generation *after* Arnold has never been the subject of a full-scale biography. Edward Thring was headmaster of Uppingham School (one of Rugby's great rivals, only a short distance away) from 1853 until 1887. He too was portrayed in a novel by one of his old boys – as Jerry Thrale, in E. W. Hornung's *Fathers of Men* (1912) – but it was a story less celebrated than *Tom Brown*. In contrast to Arnold's staff, Thring's colleagues generally stayed in Uppingham rather than seeking promotion elsewhere.

Yet, for some writers, Thring is every bit as great an educationalist as Arnold. E. B. Castle held that Thring

> was a better schoolmaster than Arnold, and allowing for the fact that Arnold inherited a large school with all its problems ready-made while Thring started with the advantages of a clean slate, his capacity for translating principles into the fabric of school life and organization far exceeded [Arnold's].[109]

Alicia Percival, who chronicled the foundation of the Headmasters' Conference, believed that 'Arnold may well be said to have changed the heart of English education, but it was Thring who changed its face'.[110] David Newsome describes Thring as 'the headmaster who most determined the shape of things to come ... his special genius lay in his realization that a school exists to educate all its pupils ...'[111] – a reference to Thring's introduction of a curriculum which ranged well beyond the purely academic, innovative specialist buildings to allow it to be delivered effectively and his passionate belief in the need to provide pupils with a safe and secure environment through a system of small, geographically distinct houses.

While any direct comparison between the two men and their reputations must wait for a future biographer of Thring, they had at least one common concern: the threat of epidemic disease in their schools. In 1841 (as we have seen), Rugby pupils had to be dispersed into reading parties in houses all over the Lake District, with others being housed in various Midland locations.[112] Three decades later, Uppingham's flight and Thring's desperate attempt to save the school which represented his life's work are the events whose background this book describes.

Existing historiography describes the Uppingham epidemic almost entirely from the viewpoint of the school – depicting the town authorities as vindictive, supine and incompetent. In the words of one writer, Geoffrey Hoyland, 'The townsfolk, though they owed their prosperity entirely to the school, were jealous of Thring's predominating influence and grudged spending a penny on improvements if they could avoid it'.[113] This view is unsurprising, given that there is no shortage of contemporary sources extolling the achievements of Thring's thirty-four year period as headmaster. These are dominated by his own writings, and especially by extracts from his diaries and letters, as well as a continuous run of volumes of the *Uppingham School Magazine* (*USM*).[114] The pro-school view which they give was reinforced by his disciples, who produced a corpus of secondary literature in the years after his death. The specific events of 1875–7 were written up in both the diaries and the *USM*, and they became enshrined in folklore shortly after the school's return, thanks to J. H. Skrine's short, romanticized account *Uppingham by the Sea*, the carefully chosen extracts which appear in Parkin's selection of documents and the other works on Thring written after 1887.[115] This folklore was reflected in a number of works over the next century, including books about Uppingham School by Bryan Matthews and Donald Leinster-Mackay at the time of the school's quatercentenary (1984) and the centenary of Thring's death (1987) respectively.[116] The latter in particular describes the Borth adventure in terms which reflect Thring's lyrical views about it.[117] Thring frequently likened the school's upheaval to the wanderings of the Israelites in the Old Testament, and referred to it as the school's 'Great Deliverance'.[118]

Other writers about Victorian boarding schools as a whole described it in similarly graphic terms: T. W. Bamford gives a racy account of Thring arriving in Borth for the first time 'on the day of a hurricane', and describes Rugby's earlier dispersal as 'a Sunday-afternoon picnic compared with the moonlight treks of the old dissenting academies, or even with Thring's epic flight from Uppingham to Borth'.[119] Alicia Percival wrote in her 1973 study of Victorian public school heads that Borth was a pivotal moment in Thring's notable career: she believed that his educational philosophy was reinforced, his critics were confounded and, as a result of the exodus to Borth, 'the headmaster's position was entirely altered'.[120]

The Victorian private school headmasters have long been known for their moral uprightness and the creative determination which the shaping of their own particular schools represented. Rather less has been written about them as organizers – and as managers of men and resources in an age before modern technology and communication systems. While the evacuation to Borth remains an act of heroic imagination, unique in its scale in the history of schools at that time, the concentration on the Welsh dimension has diverted too much of the attention of historians away from events in Uppingham itself. The traditional, hostile view of the town's RSA is too simplistic: this was acknowledged by one of Thring's successors, Martin Lloyd (headmaster of Uppingham School, 1944–65) after reading Geoffrey Hoyland's draft manuscript: 'We agreed that your account of the typhoid epidemic lets the school down rather too lightly. Not all the blame rested with the board of guardians.'[121]

This study will show the truth of that statement – although exactly how the blame should be apportioned between town and school will remain a matter of opinion. The fact that this remains the case, 125 years after the events that shook Uppingham to its foundations, exemplifies the complexity of the struggles which took place in town and school during those two momentous years.

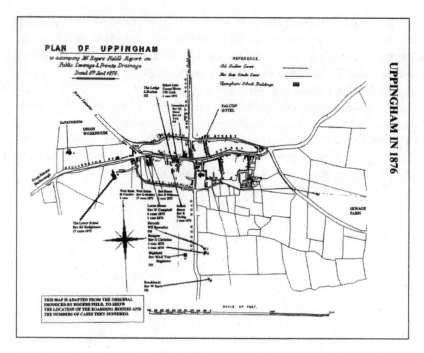

Figure 3. Uppingham in 1876, adapted from a map in A. Haviland, 'Report on the Late Outbreak of Enteric Fever in Archdeacon Johnson's School, Uppingham, Rutland: June–November 1875' (1876). By kind permission of Uppingham School.

1 TOWN AND SCHOOL, 1875

Even nowadays when the cold winds blow, people in Uppingham – 'the village of the people on the hill'[1] – are apt to state that there is no land of equivalent height between the town and the Ural mountains, thousands of miles away to the east. As recently as thirty years ago, the town was cut off by snow for an entire day, while only a small number of flakes fell in fields just a few miles away, but on lower ground. It was a small and close-knit community in 1875, similar in size to many others scattered up and down the length of rural England. Rutland, England's smallest county, is 100 miles north of London. It is only about 15 miles by 15 (95,000 acres in all),[2] and is bordered by the much larger counties of Lincolnshire, Nottinghamshire, Leicestershire and Northamptonshire. The county's entire population in the 1871 census was barely 20,000, and Uppingham was its second largest town after Oakham.[3] It had grown up around the crossroads where the north–south road from Nottingham to Northampton (each 35 miles away) meets the east–west route from Peterborough to Leicester (each about 20 miles away). Leicester was an emerging city of 60,000 people, specializing in shoe manufacture. Unlike some provincial towns, Uppingham had not undergone complete transformation before 1800.[4] Its population had grown markedly during the first half of the nineteenth century, but the increase had slowed to a halt since.[5] Nearly 1,400 at the time of the 1801 census, thereafter it grew steadily, reaching 2,065 in 1851 (of whom just over half had been born there), and 2,176 in 1861. The 1871 figures list 2,601 persons, and by 1875 the total figure is believed to have risen by about another 300 – a significant peak which would be followed by a gradual decline during the coming agricultural depression.[6] There were just under 450 inhabited dwellings.

The county boasted about 300 miles of turnpike and highway, and a small network of railway lines, but the railway had not yet reached Uppingham itself. Omnibuses departed 6 times a day – 3 from the Falcon Hotel in the market square for Manton station (3.5 miles away), and 3 from the White Hart in the High Street for the train station at Seaton (3 miles). A network of 45 local carriers provided goods and passenger links between Uppingham, its neighbouring villages, Oakham (6 miles) and Stamford (11 miles), mostly in mid-week. The daily papers did not arrive until lunchtime. Letters arrived from, and were sent off to, London and all parts of the country twice a day and once on Sundays (from the post office in High Street West). There were postal deliveries both morning and afternoon on weekdays, and a single one on Sundays.[7]

The parish contained 1,463 acres of land, most of which was enclosed under an Act of Parliament of 1770, subsequently extended.[8] Situated on a prominent ridge, the town lies significantly higher than much of the land immediately around it, and beyond it lay many miles of good agricultural and hunting country. The town itself covered about 50 acres by 1875, and consisted of one long main street (the High Street) running east–west, with the narrower North Street and South Lane running parallel on either side. These streets were joined by short lanes running at right angles to them, parallel with the Oakham to Kettering road running from north to south.

There had been a settlement in Uppingham from the sixth or seventh century (almost certainly as one of the components of Ridlington manor), although it is not mentioned in the 1086 Domesday Book land survey.[9] Most of its buildings dated from the eighteenth or nineteenth centuries when Uppingham (like Stamford) became a trading town of some fashion. Its street plan in 1875 had changed little over the previous two hundred years. Arthur Mee's *Leicestershire and Rutland* describes it as 'A pleasant little town with green fields everywhere; it has a few earthworks remaining from an ancient forest on a hill which gives us views of Beaumont Chase and the blue hills of Leicestershire'.[10] Another writer records that 'all around are green fields and hills ... dotted about are picturesque little villages in golden and brown stone'.[11]

Sandstone quarries nearby were mostly used for the buildings, although those workings which had once been close to the western edge of the town had now given way to housing.[12] Local trade directories show that the area was overwhelmingly agricultural, and that most of its population drew their income from working on the land.[13] The 1871 census, both in Uppingham and in its surrounding villages, reinforces this impression; it shows a very large number of agricultural labourers, gardeners and farm-related trades – saddlers and blacksmiths, shepherds and crop-makers.

The market had been in existence since 1281.[14] Bear and bull-baiting occurred in the marketplace just north of the church, with carts drawn up around it to make a suitable enclosure, until such events were outlawed early in the nineteenth century.[15] In the 1870s a market was held every Wednesday, and cattle fairs in March and July.[16] During these fairs, pens of sheep occupied much of the High Street; householders had to step through the pens to get on to the street, and the smells from the animals could be very pungent. Horses, cows and pigs were kept in groups all through the town. Sometimes they escaped; a cow once thrust its head through the window of one of the school classrooms in an attempt to join in,[17] and another chased a boy through part of the school, cornering and injuring him. On market day, farmers' wives rode pillion behind their husbands with baskets of produce on each arm; there was music, singing and dancing and the revellers bought hot pies and gingerbread from local street sellers.[18]

Horse racing had once taken place on an oval course on the ridge a mile to the south of the town.[19] For much of the nineteenth century there was a feast for itinerant tradesmen, held in July just after St Peter's day (29 June) – one of nearly fifty in Leicestershire and Rutland.[20] In the 1820s these had included ass racing and the ascent of fire balloons.[21] Guy Fawkes celebrations were still held; they tended to be especially boisterous and sometimes got out of hand. In 1841 some local boys were prosecuted after they let off fireworks in the street and the local constable intervened; a mob surrounded his house and smashed the windows, tearing off part of the roof. He fired on them, and several were injured. The event was stopped for a while, but it had returned to the marketplace by the early 1850s – complete with cartloads of effigies of well-known figures to be burned.[22]

Rutland has a strong continuity of names and families from the past – amongst them some still very familiar to those who now live there – Baines[23] and Cliff(e), Dorman and Ellingworth, Thorpe and Tyers. Just over half of the family names listed in the 1876 directory under members of businesses, trades and professions had also appeared in 1850.[24] This is hardly surprising in view of the evidence of the 1851 census: of the 2,065 persons listed there, just over 50 per cent were shown as having been born in the town. Moreover, of the remaining 938, half came from villages within twelve miles.[25] Out of 151 married men who had been born in the town, over 60 per cent had chosen an Uppingham woman as their bride.[26] This essential continuity gave it an oligarchic aspect; solicitor William Gilson, who was clerk to the board of guardians in 1850, changed his name to William Sheild – a name still very significant in the town in 1875.[27]

The parish church had been built in the centre of this settlement during the fourteenth century.[28] Just south-east of the crossroads, it faced the Falcon Hotel. Largely renovated and repaired in 1860 at a cost of £4,000,[29] its spire had more recently been restored and dominated the skyline. Services were held at least twice each Sunday. Closely linked to the parish church was the national school (founded in 1818),[30] which could cater for 360 children. The church's personnel was well-established; the rector (William Wales) had been in post since 1858, and of the two churchwardens, William Compton (wine merchant) had held office since 1857 and John Mould (farmer and grazier) since 1859.[31] There was an established network of church sidesmen to assist them too, including Henry Kirby (grocer), George Foster (farmer) and Thomas Bell (doctor).[32]

The trade directories list over 200 enterprises and small businesses, including nearly 30 builders, joiners, carpenters and plumbers. Of these small businesses,

28 can be classed as domestic and household services (including clock repairers and chimney sweeps). Another 35 derived their income from farming and agriculture; there were a dozen innkeepers or individuals otherwise linked to the licensed trade; and nearly 60 shopkeepers. The shops included 7 butchers and 5 bakeries. There were also 7 grocers, a greengrocer, florist, a photographic artist, and no fewer than 15 dressmakers, tailors and milliners – along with 3 doctors and surgeons and a vet. It is significant for later events that many of the shopkeepers were members not of the parish church but of one of the dissenting chapels – Wesleyan, Independent (congregational) and Baptist – or the Temperance Hall in South Street.[33]

The local newspaper, the Stamford Mercury, had been printed each Friday for more than a century. It had once been politically strongly in support of the Tory (Conservative) party, but was briefly in this period of a Liberal persuasion.[34] Its news and advertisements (no doubt read to poorer inhabitants in local alehouses)[35] give the impression of a community with a settled life. As 1875 began, Mr Tailby's hounds met in nearby Allexton at New Year, and the inhabitants of the Uppingham workhouse had just enjoyed their annual Christmas treat.[36] The Wesleyans had a public tea on Good Friday. At Easter there was a visit from a temperance lecturer, as well as a full choral service at the parish church (which launched a campaign for voluntary subscriptions for repairs to its clock).[37] Concerts there were frequent, with performers coming from all over England.[38] The Falcon Hotel (often used for dances) was about to undergo extensive alterations and the Uppingham amateur minstrels were shortly to give an entertainment in the lecture hall.[39] This had been founded in 1861 to arrange lectures for town and school alike, and next to it had recently been added a classroom, and a reading room which contained a subscription library of 1,000 books for 300 subscribers paying 1s. 6d. per year.

Theatre performances had long been held in a barn in the grounds of the Hall, just off Adderley Street.[40] The town boasted a football club and two cricket clubs,[41] and a mutual improvement society (to which reading and classrooms were 'about to be added'). The rector was the society's president; and the vice-president was John Hawthorn, who was also sub-distributor for the stamp office. He was a godly man, who ran the main bookshop and a printing business, as well as two book distribution outlets, one of which was 'the Christian Knowledge and Bible Society Depot'. J. C. Guy (manager of the Uppingham branch of the Stamford and Spalding bank, one of three banks in the town) was its secretary, and the treasurer was Charles White (an ironmonger on High Street East).[42] There had also been a book club, a subscription reading society and 'news room, well supplied with the London and provincial journals etc'.[43]

Local government officials in the town included a registrar of births, marriages and deaths, and an Inland Revenue officer (based at the Falcon Hotel). For those who fell foul of the law there was a county court held every two months at the Falcon, with jurisdiction over surrounding villages. Four local magistrates took turns to sit on the first Friday in each month.[44]

The two law firms were involved in a substantial range of activities beyond purely legal work. First and foremost they were stewards for the two local manors. The family practice of William, William Thomas and Robert Sheild (with John Pateman in partnership) acted for the manor of Preston and Uppingham, held by the Noel family. William Sheild had worked in the town for three decades but lived in Wing, where he had inherited property.[45] He was clerk to the magistrates and to the county courts, and not only a Poor Law guardian but also a solicitor and superintendent registrar of the Uppingham union. He and his partners held a string of local posts as variously clerk to the justices, commissioner, registrar, bailiff, coroner, treasurer and turnpike trustee. William Thomas Sheild and John Pateman were between them agents for no fewer than five insurance companies. William Henry Brown acted for the Rectory manor. He came from a long-standing Uppingham family, and was the third generation of his firm to work in the town.[46] He was clerk to the guardians and the RSA, as well as clerk to the workhouse and commissioner to the supreme court of judicature. Both law firms lent money and carried out extensive property transactions, which also involved arranging mortgages for clients, many of whom ran shops and small businesses.[47] Both William Sheild and W. H. Brown were also board members of the Uppingham Gas Company.[48] They were men with knowledge of their clients' affairs, local influence and considerable local patronage – as was bank manager Guy, who was agent to a further four insurance providers.

There was a fire station on the Glaston Road and a police station on Stockerston Road (with two officers and two cells). After an initial proposal to install gas lighting in 1831 (rejected as requiring too heavy a burden on the rates), concerns about safety and security led to a second, successful scheme in 1839, when a gas works was built on Stockerston Road.[49] The system was further improved by new street lights with 'handsome cast-iron pillars' in 1860, but supplies in both houses and streets were not always reliable.[50] Gas was used mainly for lighting until well into the twentieth century – although the cost and effectiveness of this and other services were the subject of dispute between the vestry, town members and the rector over a long period.[51] There would be no electricity for many years yet.[52]

Between a third and half of the population of England and Wales lived in, or was dependent on, provincial market towns in 1851.[53] Rutland itself would change comparatively little over the next half-century in this respect.[54] Uppingham was typical of much of rural England: stability and continuity were evident everywhere within it.

What was Uppingham's sanitary state in 1875? The High Street had been bordered on both sides by open channels of water until the 1840s – streams which turned ashy on Mondays (thanks to blue-bag and soap as people did their washing in cauldrons or coppers supported by bricks over wood fires, before leaving it spread out to dry on bushes or grass) and olive-green on Wednesdays thanks to the gathering of sheep and horses at the market.[55] 'Wednesday's water, *horribile dictu*, was after filtering used extensively for brewing beer. Of the colour of weak coffee and with a strong odour, it was much esteemed for this purpose.'[56] It had been common practice for waste of all sorts to be thrown into pits behind dwellings.[57] The earliest privies (outdoor toilets) were built over these ash-pits, later filled in and replaced with large buckets. Many of Uppingham's cobbled streets were still largely dirty and ill-drained: in 1866 some were replaced two centuries after they had first been laid down, and generations of Uppingham schoolboys remembered the 'obnoxious cobbles' of the High Street.[58] These problems had been exacerbated by the town's growing population, which had to be housed in limited space; over time, building had tended to take place not on former open fields now enclosed, but behind the existing street-front houses and shops. Gardens had been turned into yards as cottages were erected in small spaces – often with restricted access through existing buildings or down short lanes. The wealthier citizens tended to move to the edge of the town, while the poor became concentrated in certain areas such as Stockerston Road and Adderley Street.[59] In 1819 the Rectory manor court heard a complaint of nuisances in the wash pond 'by animals being thrown therein and the mud suffered to remain too long'.[60]

In this era there were some areas of particular concern. One was Ragman's Row (or Rag Row) – a group of cottages and shacks off North Street West, where 'the houses were of the poorest – mud walls with very low doors and unglazed windows'; 8 small dwellings housing 36 people including (amongst others) several paupers and a scavenger, a washerwoman, a laundress and a pauper shepherd. The acquisition by the school of many of the properties along the south side of High Street West drove families into places such as Deans Terrace and Wades Yard; many poorer inhabitants were housed in back yards. Innocents Yard had a density of occupation of 122 persons per acre.[61]

After centuries during which sewage and rainwater had flowed more or less unchecked (apart from a rudimentary and small-scale system of wooden drains installed around 1786),[62] a main sewer had been laid through the northern part of the town along the High Street and North Street in 1857–8. But it had been built only four or five feet deep, with stoneware pipes of 12 and 15 inch diameter,[63] and the system had never fully been completed. A new, deeper main drain, known locally as 'the new south sewer', had been constructed along South Lane in 1872, connecting with the older one to the east and running along Stockerston Road to the west of the High Street, with branch extensions at various points. A sewage outfall works had been added in 1874–5 near Seaton Lane, to the south-east of town. As a system to serve a growing town and school, however, it was seriously inadequate.

House drainage was frequently unconnected to these new drains – which prompted sanitary expert Rogers Field, reporting to the RSA, to describe Uppingham as 'still a cesspool town'.[64] He could not examine every house, but he estimated that three-quarters of them drained into cesspits, and of 379 houses visited, only 84 drained into the sewer system. There were 174 houses with water closets which still drained straight into cesspools, often in gardens, and 34 houses which shared exit pipes with next-door properties (with the consequent risk that waste might back up into a neighbour's drains or even into the closet itself). Of the remainder, 6 had pail and earth closets in the house, and 81 had no indoor facilities, but used outside privies in which the waste was mixed either with water or with dry earth and ash. Many of these were, he believed, inadequate or broken down.

There was a second potential problem: in 1875 Uppingham still had no waterworks. It relied on well-water (public or private) for drinking, and for servicing any water-closets. The geology of the area suggested that the well water ought to be healthy. The land around the town had long ridges of low but steep hills separated by fertile valleys,[65] thus channelling surface water effectively into streams supplying wells in the basements or gardens of many local properties – including the boarding houses owned by the school. A small tributary of the river Welland flowed through the town.[66] On its south-east side, Uppingham was flanked by Lincolnshire limestone (richly fossil-bearing),[67] which stretches from Northamptonshire into Lincolnshire – thought to be generally healthy as a water source. To the north-west, anyone digging through the upper lias clay and the Northampton sand ironstone (known locally as the 'kale') to a depth of twenty to thirty feet would reach the blue lias clay and a plentiful water supply. One observer noted:

> The soil is of red appearance. Beneath, to the depth generally of two or three feet, is a shady red stone, and under this as far as it has been worked, either a red stone, or a blue stone encrusted with red, of variable thickness, and a very stiff blue clay which makes good bricks. The red stone is soft and easily worked; the blue is much harder: both are used for building.[68]

Theophilus Rowe, who lived on the west part of the High Street in 1875, drew on his memories twenty-five years later (in 1900, by which time the Uppingham epidemic had taken place and bacteriology had moved on apace),[69] to give a less flattering view:

> Dig down anywhere in Uppingham for 14 feet through the kale, and then 4 feet further into the lias clay, and you have on your own premises a draw-well, not too deep, with a water-holding cistern at the bottom. But the water? The kale no doubt is an excellent filter; in ordinary times its loose joints aerate the drops as they trickle through, and the iron with which it is tinged disinfects them. And in fact there have been ... many long periods during which this filtration has sufficed, and sanitary statistics have not claimed attention.
>
> Uppingham was always by tradition a healthy place, rather priding itself on its bracing breezes and plentiful springs. But can we doubt that from time to time within the town area not only the pure rain-drops from the sky but many drops not so pure filtered through the loose kale, and that in those sparkling wells from time to time there sported millions of bacteria, enough to account for whole consorts of fevers?[70]

There had been a scheme as early as 1826 for supplying water 'from a fountain above the town ... The spring throws up without intermission a very large supply of excellent water, well situated for being conveyed by pipes ...'.[71] Little seems to have come from it however, and half a century later those who had no wells of their own had to trundle water-carts or carry buckets to public supply points from a series of springs all over Uppingham.[72] The town spring, to the south of the Stockerston Road at the west end of the town, was particularly heavily used; its feeders passed quite near to the Lower School.[73] There were two other public wells, one in the marketplace (erected in 1818) and the other a short distance away outside the front of the Methodist chapel.[74] In addition there were pumps in many of the courts and yards and 'a fine stone drinking trough' at the bottom of Leamington Terrace.[75] Many properties had their private water supply, from springs, wells and ponds. Deeds often recorded rights of access to a pump in a neighbour's yard.[76] At the beginning of the century the innkeeper of the Chequers inn had laid an unauthorized pipe from a public spring adjoining the horse pond into her premises.[77] It is likely that many of the inns brewed their own beer. In addition, no amount of sound geology could make up for pollution – especially on the western side of the town, where most of the school houses were situated. Manure heaps were one source of it; for example, the heap in Reeves Yard in 1863 was right next to a pump.[78] 'Uppingham was unkempt; its houses were tumbledown, yards untidy, heaps unfenced; it was unpleasing to the eye and unsavoury to the nose'.[79]

In one respect Uppingham was an unusual town. Within its 3,000 or so population there was a school sub-community comprising nearly 15 per cent of its total, in term-time at least. It had once been a small Elizabethan grammar school, and it had had barely a dozen pupils as recently as 1839.[80] The Elizabethan schoolroom, the original building from its foundation in 1584, lay just below the church to the south-east. Over the previous twenty-five years the school had been re-founded by its headmaster, Revd Edward Thring. He had created a boarding community of over 300 boys, together with well over 100 adults (masters and their families and house servants) who occupied over a dozen boarding houses.[81] Some of these houses were Uppingham's newest and largest properties. By 1875 the growth of the school had put severe pressure on the town's public services – or lack of them.

Town and school personnel interconnected with each other in a number of different ways. The school invited townspeople to some of its events; there was a school concert in March 1875 featuring Josef Joachim, the leading violinist of his day.[82] School pupils regularly gave concerts and dramatic readings for townspeople; there had been a performance of Sheridan's *School for Scandal* in the schoolroom in December 1874.[83] Thring was keen to foster links between school and town, and was aware that the school had much better facilities than the town; he had proposed establishing a recreation ground on the Leicester Road, with a bowling green and archery and croquet facilities.[84] The school gave an annual Christmas party for children from the workhouse situated on the Leicester Road a few hundred yards away. There was also the day-to-day movement of school personnel between different parts of the town (as Figure 3 shows). Unlike some of the major boarding schools founded in the mid-nineteenth century, Uppingham School was a dispersed community, rather than a school built on a single campus. The houses (with their own individual characters and gardens) and other school buildings were mostly situated along the western part of the High Street, but spread out for some distance along the main east–west road through the town. There were four other houses about half a mile from the rest, on the hillside to the south.[85] This resulted in continuous daily contact between town and school as pupils and masters went to and from houses and lessons (taught by housemasters in their house halls and by other staff in makeshift classrooms and laboratories all over the town), or to visit their friends in other houses, or as they took part in afternoon races and steeplechases and paper-chases along the surrounding roads and fields; following hounds or exploring nearby Stoke and Wardley woods.[86] According to one diary, housemasters' wives ran the domestic side of the houses (including the catering arrangements), sending out and accepting dinner invitations from other houses, corresponding with parents, walking the dogs, attending chapel and watching cricket matches or athletics races.[87] All this activity involved a large amount of contact (and friendships) with people in the town. Some staff appear to have been conscious that relying solely on the school community for their social life would have risked becoming introverted and claustrophobic. A housemaster observed on one occasion: 'As masters we are admirable, but as men we vegetate'.[88] It may be no coincidence that he had grown up as a day-boy in London rather than as a rural boarding pupil.[89]

There were occasional tensions too. With so many boisterous schoolboys living in this small town alongside its other inhabitants and sharing its narrow streets, coexistence was not always easy, and it is hardly surprising that relationships sometimes became strained.[90] Thring, complained in his diary in 1860:

> A stupid complaint from Lord Berners again about the boys being in his wood. I don't think he ever can have been at a public school, and he is bitten with the curse of the English squirearchy, the pheasant mania.[91]

A former pupil remembered:

> Some pupils preferred to go round in groups after dark ... In the [eighteen] fifties there was no love lost between town and school, and a collision with 'the cads' or a chevy from [chase by] a labourer in the fields was a common thing. We were always easy to tell by our caps ... The worst time, and as a rule we brought it on ourselves, was in winter when there was snow. Then a 'fight with the cads' was the proper constitutional thing, and we fought with some fury, more especially when we found that they were not fighting fair, but were putting stones into their snowballs. They generally lined the churchyard railings, and fired at us from above as we came from the school 'quad' or the market-place ...[92]

The same writer paints a revealing picture of a town which was wary of the school:

> It would not be unnatural to suppose that the town was proud of its school; or at all events, as it grew in numbers, and trade of all kinds kept increasing, that the tradesmen would recognize that it at least had its uses. But it was not so. The farmers and publicans spoke of 'them dratted scholars', and the tradesmen would, with few exceptions, say that the school was nothing to them, and that they were best without it.[93]

On the other hand, even though fraternization of pupils with townspeople was not encouraged, some pupils made friends of those living in School Lane and Leamington Terrace. They also managed to sneak into market day and the two annual fairs.[94]

Finally, the school must have had a significant impact on the town's economic fortunes in two ways. The town had a large number of shops and small businesses in relation to its size and population, suggesting that there were plenty of people in the town who relied on the school and its pupils as customers.[95] Moreover, given that the pupils' parents could afford the school's sizeable fees, their sons' collective spending power must have been very large when compared with that of most of Uppingham's townspeople. The bakeries sold plum shuttles (pronounced *shittles*), a type of bun especially popular with children[96] and, no doubt, with the pupils at the school, who particularly liked the hot rolls produced in the mornings by two local bakers, Loves and Baines. Henry Kirby, the grocer in the marketplace, sold ice creams and lemonade to boys during their free time in the afternoons, as well as strawberries and cakes; the editor of the *USM* complained that the large amount of time that they whiled away in Kirby's establishment could have been better spent back in the school, writing articles for his publication.[97]

John Hawthorn at the post office and bookshop was another shopkeeper well known to the boys: 'garbed in somewhat clerical clothes, with rather short trousers displaying white socks, a clerical hat and a monocle attached to a broad black silk ribbon. Well might he have featured in a Trollope novel ... a distinctly awe-inspiring figure.'[98] The boys who stocked up their supplies of stationery from his bookshop had a strong affection for him and nicknamed him 'Sempy', short for 'sempiternal' (or eternal); he was a man 'always of the same youthful manner and appearance' who – it was rumoured – had been in the town for ever, and who traded there for nearly half a century.[99] Not surprisingly, he was one of the school's strongest supporters.

The decentralized nature of the school increased the number of local suppliers whose business well-being depended on a successful and healthy school in every sense. Because of the dispersed houses, with pupils feeding in their own house dining rooms, and with no central catering or purchasing system, each housemaster and his wife made their own decisions about suppliers. Hawthorn also had town loyalties and ties, however: not only did he supply the school with books and print its exam papers, but he was also a supplier to the rector and the church, and he provided their service sheets.[100] Grocer Henry Kirby was another tradesman with many connections in the school.[101]

The school was also a large-scale employer. Its pupils and staff all had to be fed, accommodated, and provided for in a variety of ways. The houses employed nearly 100 living-in staff between them; the 1871 census showed that four of them had between 7 and 9 each – including various governesses, a few footmen, numerous cooks, nurses, parlour-maids and kitchen-maids and one 'boots'.[102] To these would have been added a large army of people living in the town but working in the school by day – self-employed, or on piece-work.[103] Houses had to be repaired and altered; some were still being built or developed. Furniture and equipment had to be ordered and maintained, and gardens tended. In an age when few parents visited the school, boys had to be clothed, and their clothes needed cleaning and repairing.

Farm produce would have been purchased locally by the school. The food shops in particular must have noticed a big drop in their turnover when the holidays began. All in all, town and school were highly interdependent economically. the school would suffer in reputation and well-being if local businesses failed. For those businesses the presence and goodwill of the school was a key factor in their continuing prosperity and development.

The interlocking set of social and economic relationships between town and school can be demonstrated by studying the 1871 census returns for the High Street. Within less than half a mile there were a number of boarding houses, including those of Edward Little (with his wife, child and a sister-in-law, plus 4 domestic servants and 14 boarders), Sam Haslam (with his wife, 6 servants and 30 boarders), Revd Walter Earle (with his wife and family of 5 small children, 8 domestic servants and 27 schoolboys) and William Campbell (with his wife, 7 children, a sister/governess, 8 servants, 4 overnight visitors and 33 boarders). Across the road were the houses of Theophilus Rowe (with his wife, 4 servants and 31 boarders) and Bennett Hesketh Williams (with his wife, 5 children and 2 other relatives staying, 6 servants and 16 boarders). Their more well-to do neighbours included professional people: Guy the bank manager, Bell the doctor, Pateman the solicitor, and others such as master butcher and multiple shopkeeper Peter Fryer and successful farmers William Mould and John Shield.[104]

By no means all the neighbours were so prosperous. Sandwiched in between these professional men lived a network of small businessmen, traders and artisans – including a master boot-maker, a saddler, a shoemaker, an auctioneer-cum-estate agent, a master watchmaker, a chair-maker, a laundress, a grocer, 2 drapers, a hairdresser and an innkeeper. Further up the street were a mason (married to a dressmaker), a railway agent, another draper, a tailor with his wife and 4 children, and a plumber and painter (with his wife and 5 children) who lived cheek by jowl with one of the curates. With so many trades, goods and services represented amongst their neighbours, it seems likely that the housemasters' personal and business relationships would have overlapped to a sizeable degree.[105]

Those running these boarding houses included Revd Edward Thring who, by 1875, had been headmaster of Uppingham School for 22 years. Born in 1821 in Somerset, he was the third son of a Somerset country gentleman and rector. One of his brothers, Henry, became a notable parliamentary draftsman, and was knighted in 1873.[106] After Eton, Thring progressed to King's College, Cambridge, and was ordained as a deacon in the Church of England in 1846, serving a curacy in a run-down area of Gloucester – a difficult time which included some elementary teaching. However, he suffered a breakdown within a few months, and, after travel in Europe and a whirlwind romance in Germany which led to his marriage, he returned to England and, despite the scepticism of his family and his limited experience of working in schools, he was appointed to be headmaster of Uppingham in September 1853.[107]

Thring formed distinctive ideals about education, developed over 35 years in a whole series of writings, as well as published sermons. An academically average boy should have as much time and money spent on him in the classroom as a brilliant scholar – in contrast to the views of the famous Dr Thomas Arnold of Rugby School a generation earlier, whose dictum (Thring believed) had been that 'the first, second and third duty of a schoolmaster is to get rid of unpromising subjects'.[108] Classes would be allocated to staff according to their particular teaching talents rather than their seniority in the hierarchy: 'to teach an upper class requires more knowledge, a lower more skill as a teacher'.[109] A good school needed good extra-curricular facilities – 'machinery'.

Archdeacon Robert Johnson had endowed schools and almshouses in Uppingham and Oakham in 1584,[110] but the schools had remained quite small for over two centuries thereafter. Thring's arrival coincided with a great expansion in middle-class education as the Victorian industrial boom began; significantly for future events, a number of his pupils' parents were doctors.[111] Thanks in part to the growth of railways, between 1853 and 1875 Uppingham acquired a national catchment of boarders, but the local places for day-boys largely faded out.[112] Thring's original 43 pupils grew to 100 within 6 years, doubled again within the next 4 and had reached his chosen ceiling of 300 in 1865. We know where roughly 95 per cent of the pupils lived in 1875. Apart from ten living abroad, they came from all over Great Britain and Ireland. Barely a dozen of them were from Uppingham or the surrounding villages; the school had moved well away from its local, free grammar school roots.[113] Liverpool/Manchester and Lancashire accounted for nearly 70 of them, and another 42 are listed in the *USR* as coming from London – two areas of the country which had recently enjoyed the most advanced public health arrangements under Sir John Simon and W. H. Duncan respectively.[114]

By 1872 Thring had 21 teaching staff – a big running cost, but one which he believed to be essential.[115] He also decreed that 23 should be the optimum size for a class and 31 for a boarding house.[116] Not surprisingly, a number of his staff doubted his arguments for such limits, especially in times when budgets were tight, and by 1875, 3 of the 12 houses had crept up to 37 or 38 pupils; most of the others (including Thring's own) were between 30 and 34 and the rest were between 20 and 30.[117] He and 11 of his staff were housemasters. They brought very different but complementary temperaments and capabilities to the enterprise – firm discipline, innate caution and love of cricket (William Earle of Brooklands, the usher and longest-standing member of staff, who had joined the school in 1849), financial acumen and solid gravitas (Campbell of Lorne House), organizational skills (Candler of West Bank, the mathematician/timetabler), scholarship (Sam Haslam of The Lodge), an 'unbending demand for work' (Charles Cobb, of the house on the corner of School Lane), and eccentricity, other-worldly godliness and calm resolution (George Christian, housemaster of Redgate and school chaplain).[118]

There was also Revd R. J. Hodgkinson at the Lower (junior) School – a legally and financially separate institution from Uppingham School itself, but one which sent many pupils on to Thring for their later education. Unlike the houses in many of the boarding schools founded in the nineteenth century, Uppingham's boarding facilities were not to be found merely in a series of buildings on a single campus served by a communal dining hall. They were physically separate from each other and from the central school buildings, individually designed to reflect the wishes of the housemasters who built them, each with their own staff and house dining hall. It made each of them – and their ethos – varied and distinctive. It may also have contributed to a plumbing system of uneven quality with little central control: boys washed in the mornings in chilly stone-floored washrooms with rows of stone basins filled with water from cisterns which took up to two hundred strokes of the pump serving them.[119]

The housemasters were mostly men of at least moderate private means and thus able to afford to commission their own architects and build their own houses.[120] Some converted an existing house in the town or bought one which was already a going concern. A few started off in a small town house and then built a larger one on the outskirts.[121] Unlike modern counterparts, they had a direct financial stake in the school. Several appear to have taken out large mortgages: George Mullins started off in Red House, but then paid £3,100 to buy West Deyne from his predecessor, and took out a loan of £2,000 to do so.[122] They would eventually sell these houses on to their successors. Meanwhile their investment would be at risk if the school were to fail – and in a small country town there might well be a shortage of alternative uses for (and buyers of) large properties if things went badly.

As a housemaster himself, Thring had a common financial interest with his colleagues in the collective health and strength of the school. If his staff had risked much in terms of their capital investment, he had committed even more himself. He was always in debt – as early as 1857 to the tune of over £2,750,[123] through guaranteeing and paying the salaries of his growing staff. He was forced to take out a loan of nearly £2,000 over fourteen years from the Wellingborough Building Society; it concerned him greatly and wore him down, but he tried to keep these money worries from his wife and children.[124] He also had one other very direct instrument of control. In the 1870s, parents paid boarding fees to the housemaster, and only tuition fees to the school itself.[125] Housemasters made profits (or losses) from their houses, and derived most of their income from being housemasters. They were paid comparatively little in fixed salary as classroom teachers. Thus they relied on Thring's recommendation;[126] a housemaster judged by him to be inadequate to the task could soon be starved of prospective parents. They had to conform to his standards of food and accommodation as well as supervision and care. If they pared away at standards in order to increase their profits they could be rapidly frozen out, and he was resolute that he would not let them increase their numbers purely to increase their revenue.[127]

After some early appointments which Thring came to regret,[128] the housemasters of the 1870s were a more settled group. They were nearly all graduates of

Oxford or Cambridge and mostly from professional families, although few had any background in teaching. Several would run their houses for over 30 years.[129] Besides Thring himself, 10 were in holy orders. Nearly all had a wife; Thring, happily married with 5 children of his own, regarded the part which the house-master's wife played in each house as one of its most humanizing influences.[130]

The school's scheme of management had been revised as a result of the Endowed Schools Act of 1869 and the resultant Taunton Commission. Thring explained the school's financial arrangements in some detail in 1880:

> Each boy pays £70 per annum to the housemaster. This sum does not pass through the hands of the governing body. Each boy pays £40 per annum as tuition fee [up from £30 in 1875]. This sum does pass through the hands of the Governing body, and is now distributed as follows: One sixth, £6.15s to the headmaster, fixed by law as the lowest proportion; £24. 18s per boy is assigned for the payment of masters up to the number of 320 boys; £1. 10s per boy for current expenses; £6. 15s per boy for reserve fund. £100 per annum is assigned for lectures, readings, concerts etc; £100 for prizes in the school. There is also a sum of £1,100 per annum from the original foundation, which is mainly expended on providing exhibitions from the school to any university or place of higher education, of the value of £60, £50 and £40, three every year, tenure for three years, and in a small salary of £200 per year for the headmaster.[131]

Thus Thring and the housemasters effectively had control of £70 for each of the 300 or so pupils – around £21,000, out of which they had to pay all the boarding expenses, after which they took their profits or losses, and decided whether to contribute to future building and staffing projects. The governing body controlled £30 per pupil in 1875 (*c.* £9,000), together with the income of the Archdeacon Johnson charity set up by the founder 300 years earlier. In the early 1870s this was drawn largely from rental income on lands in Warwickshire and Lincolnshire, and amounted to £4,280, of which ⁴⁄₇ was spent roughly equally on the two schools in Uppingham and Oakham, and the remaining ³⁄₇ was specifically designated for the almshouses.[132] It is clear from Thring's statement about the cost of education at a good boarding school, made to the commissioners in 1866, that he felt that Uppingham's fees were barely adequate for the efficient running of his school.[133]

Thring and his housemasters did not finance just the boarding houses. As the school grew, they subscribed to many building projects which the trustees were unable or unwilling to finance. On his arrival in 1853 the masters had given £500 towards the restoration of the parish church, which the school used daily in the early years. However, by 1858 he was commissioning a leading architect of the day, G. E. Street, to draw up plans for the school's own chapel, and for a new schoolroom for assemblies. The trustees reluctantly agreed to borrow money to pay for the schoolroom, but the £10,000 for the chapel was largely financed through a general subscription from masters, parents and friends. It was completed in 1872.[134] These buildings were constructed to a high specification and were substantial in size. Simultaneous expense was incurred through a similarly ambitious gymnasium. In 1859 Thring had persuaded the masters to buy the Cross Keys inn for £1,130, just to the west of his own boarding house; an unexpected opportunity to acquire a key central site for future development.[135] This too was largely funded by the masters, who were simultaneously contributing to several other projects.[136]

The result of all these building projects was that the masters had financed an increasing proportion of the school's development. In the mid-1860s, over 90 per cent of the school's buildings, land and equipment had been financed by Thring and his staff, while the trust had provided a mere 8.75 per cent.[137] By 1875 the masters had spent over £40,000 on buildings.[138] The school was prospering, but if times were to change for any reason they would all have plenty at risk. Thring had the most to lose of all.[139] There was also the potential for dispute between the masters and the governing body in any time of economic or other difficulty, given that the personal finances of Thring and the masters were so inextricably bound up with those of the school itself.

Thring was extrovert and enthusiastic – at times impulsive. He had a brain which moved in intuitive leaps, which drove a passionate and impulsive nature. He was also committed to spiritual simplicity; in contrast to the high-church rector, of Uppingham's parish church, William Wales, he had little time for ritualism or doctrinal minutiae.[140] It was unlikely that the two men would ever warm to each other. Thring would also fight tenaciously to protect his school if anything threatened it: it represented both his livelihood and his life's work. In 1875 he was also just emerging from a period of prolonged battles to protect the school externally – against the attempts of the government-appointed Endowed Schools' Commissioners to restrict the independence of schools and their headmasters. One by-product of this struggle had been the creation of the Headmasters' Conference (HMC), whose first meeting had been held in Uppingham in 1869.[141] As two headmasters of famous southern public schools travelled by train across the sodden countryside of the East Midlands to Uppingham for that first conference, one of them observed to the other: 'Thring must be a wonderful man to have made a school like this in the midst of such a howling wilderness'.[142] It was a harsh verdict, from a man perhaps missing the familiar leafy lanes of Kent.

2 LOCAL SOCIETY AND LOCAL GOVERNMENT

Uppingham was situated in a county which was a good deal more than a howling wilderness. In its structure and outlook, Rutland was quintessentially rural: it might even be described as still feudal. Of its 96,000 acres, nearly 79,000 (82 per cent) were under cultivation of some sort – with 42,000 acres given over to arable and 36,000 to pasture.[1] Landed interests predominated everywhere, and landowners were unlikely to be sympathetic to rapidly rising rates.

The influence of the leading members of the Rutland gentry was exercised largely through the ownership of land and property. Walford's list for 1876 names 35 families who represented the core of Rutland society,[2] and although its diminutive size makes statistical comparisons with other counties suspect, an indication of Rutland's character is seen in the fact that it had one country house for every 31,000 acres – the highest such distribution of any English county.[3] Four of the 331 greater landowners in the modern Domesday Book (the so-called blue book) of land ownership, drawn up by John Bateman in 1873, had 9,000 acres or more in Rutland, and between them they owned half the county. Two of them had their chief seats in Rutland, and the other two were just across the border in Lincolnshire. Apart from Rutland, only Northumberland as a county had more than half its acreage in the hands of owners of 10,000 acres or more.[4]

Bateman also records the names of owners of land of more than 3,000 acres with a gross annual value of over £3,000. Of Rutland's total acreage, over 70 per cent (66,294 acres) was owned by one peer, five 'great landowners', and five squires. No fewer than four of this elite group were trustees (governors) of Thring's school: the Earl of Gainsborough (who lived at Exton Park),[5] Sir John Fludyer (Ayston Hall), George Finch MP (Burley-on-the-Hill) and John Wingfield (Tickencote Hall). Below them, a further 10,017 acres were owned by 'great and lesser yeomen' or medium-sized farmers. Compared with other counties, a large proportion of this group was made up of clergymen.[6]

By contrast, at the other end of the scale there was an unusually high proportion of small landholders.[7] The average size of holding in the county (14.75 acres) was the fourth-lowest of the 54 counties in England and Wales which

Bateman analysed. There were 458 'small proprietors' holding a total of 6,782 acres, and 132 acres were held by no fewer than 861 cottagers, while 53 public bodies accounted for a further 2,392 acres. The remaining 401 acres were described as 'waste'.

The land tax assessment for 1874/5 shows that 258 people were assessed, of whom three stand out far above the rest.[8] Between them they accounted for over 45 per cent of the total assessment. Not surprisingly, one was Lord Gainsborough,[9] who owned two houses and lands on Beaumont Chase. He was a quintessential county landowner – once High Sheriff and subsequently Lord Lieutenant, as well as being a lieutenant-colonel in the Leicestershire yeomanry and (briefly, in 1840–1) a Whig MP.[10] He had large estates in two other counties,[11] and he owned lands in no fewer than 19 local towns and villages, including nearly all the property in six of them. His total landholding in Rutland was 15,000 acres, with another 3,500 in five other counties. His family (the Noels) had long held the Manor of Uppingham and Preston, and it was said that the Earl could ride from Exton to Uppingham (nearly ten miles) without going off his land.[12] The second of these large landowners, Sir Charles B. Adderley MP, was one of Bateman's 'outsiders'. He had inherited his lands from his uncle.[13] Like Gainsborough, his lands were let out to tenants; his main estates were in Staffordshire and Warwickshire. He was an uncompromising Tory who had opposed Peel on free trade. He had been briefly President of the Board of Health in the Conservative government of 1858–9, and Under-Secretary for the colonies in the administrations of Lord Derby and then Disraeli in the late 1860s. From 1874 he was President of the Board of Trade (1874–8) before going to the House of Lords. He was no great speaker, but he was a 'capable business administrator and he could make a plain business statement very well'.[14] He never lived in Uppingham, preferring to put tenants into the Hall in High Street East. Both Adderley and Gainsborough held buildings and land assessed for rating purposes at over £18. The third of this trio came a long way behind the first two, but his presence is significant. The rector William Wales was a school trustee *and* a town guardian, as well as being chancellor of the Peterborough diocese.[15] Another 18.5 per cent of the land in Rutland was in the hands of ten further owners, who included two other absentees (William Belgrave and Lord Aveland). The remaining 36 per cent was shared between 216 individuals.

The return of owners of land shows that Wales's land-ownership yielded a gross estimated rent income of no less than £747.17.0.[16] He enjoyed rents from those leasing his glebe land, manorial rents and fines from his copyhold tenants, pensions (in lieu of former tithes) amounting to £6.14.4 per year, and the annual Easter church collection.[17] The Rectory manor was made up of land which mostly lay to the west of the Oakham–Kettering road and included much of the area on which the school and its houses now stood. It was smaller than the

Preston and Uppingham manor, but it enjoyed one significant advantage over its neighbour: the Rectory manor's entry fines were arbitrary rather than fixed.[18] Both manors had once held courts once or twice a year, although these had now fallen into disuse and their powers had been transferred to stewards who were also local solicitors; William Sheild for the Noels and William H. Brown for the rector. Sheild was a significant copyhold and freehold property owner in his own right and was also a moneylender. Sheild and Brown had both attended the school; Brown was one of six sons of Thomas Brown, solicitor, to do so.[19]

In Uppingham itself the 1873 return of owners of land produces a similarly revealing picture.[20] Five men owned more than 100 acres: Conant, Fludyer, Wales (all trustees) and two town guardians, John Parker (a farmer from the nearby village of Preston) and William Sheild. Of the other guardians, two – George Foster (a farmer, but also a property-owning solicitor) and Edmund Robinson (a dealer in corn, glassware and china) – appear in the list of four owners in the 50–100 acre category. William Mould (a farmer and maltster in the High Street) was one of five in the 20–50 acres list, and William Satchell (a builder) appears in the list of others – along with the guardians' clerk, W. H. Brown, Thring and five of his staff. The guardians themselves, the school trustees and the churchwardens also appear here as institutional owners.

The list of owners of houses and buildings is again dominated by Gainsborough and Adderley – although guardian William Mould also appears in the highest category. Below them it is possible to identify three groups in addition to retired people or those with private means. One is the professionals: one surgeon, three solicitors and two bank managers. The second is the school's masters who, with the school itself, collectively contributed between a fifth and a sixth of the rates collected.[21] The third is the shopkeepers, of whom at least twenty appear in the top third of the total list. They cover a wide range of businesses, including those of draper, ironmonger, grocer, bookseller, chemist and hairdresser, and many must have been suppliers of goods and services to the school as well as to people in the town – trade for which many of them would have been in competition with each other. The assessment does not, however, yield information on two issues: the amount of precarious mortgage debt which many of the smaller owners in particular must have carried, or the extent to which landowners passed on rate increases to their tenants.[22] Most of the shopkeepers listed in the houses and buildings register were owners, but not all.

The landed influence was strong within the board of school trustees, who by 1875 were 19 in number. One was the hereditary chairman: 'the right male heir of the Founder', Mr A. C. Johnson. He, along with Gainsborough, Conant, Finch, Fludyer, Watson, Wingfield and Wales, has already been identified as a significant property holder.[23] The representative trustees also included Rt Hon. Gerard J. Noel, the Earl of Gainsborough's second son, a Conservative MP and former junior government minister, from Catmose Park, Oakham.[24] A number held office ex officio (i.e. by virtue of another office which they held, and were thus unelected) – Gainsborough was a trustee by virtue of being the Lord Lieutenant of Rutland, and Fludyer through his chairmanship of the county quarter sessions. Not all the trustees were landed gentry, however.[25] Thring and the masters nominated two of them; and it is no coincidence that this pair came from a quite different, business background. Both had sons at the school; both had been members of a parent group that had rallied to Thring's support against the Schools Inquiry Commissioners a decade earlier,[26] and both lived far away – in that part of the country which had the strongest concentration of pupils, the industrial north-west of England. Thomas Birley from Pendleton, Manchester, was a cloth manufacturer; Wensley Jacob from Birkenhead (across the River Mersey from Liverpool) was a merchant and local JP. The clerk to the trustees was another local property owner, although only on a very small scale, J. C. Guy, the bank manager.[27]

Of the sixteen governors who held office during Thring's critical early years at the school, all lived within 25 miles of Uppingham, and 13 within Rutland. Six were local clergy, nine were squires 'living in large houses in the villages they mostly owned',[28] and one was both (like Thring's own father). Five had been high sheriff of their county; no fewer than nine of them were local magistrates.[29] Their average age in 1855 was just over 65. It was with the passing of the Endowed Schools' Act in 1869 that the governors acquired the new title of *trustees*, and their number grew.[30] They became an abler group overall, but little had changed in their dominant economic preoccupations – or their narrowness of outlook and vision. Even in the 1890s, Mandell Creighton, who was a trustee himself as Bishop of Peterborough and a far keener observer of local economic affairs than many bishops,[31] would admit to one of Thring's housemasters: 'There are several bad governing bodies in England, but none nearly so bad as ours'.[32]

Thring had never enjoyed an easy relationship with the trustees. They were the guarantors of the funds of the Archdeacon Johnson foundation, responsible not just for Uppingham but also for its sister school at Oakham. Thring believed that they were out of touch; 'mean-spirited consequential dignitaries'.[33] They were sensitive to criticism within the town that the school no longer catered for the aspiring poor of the local community, but was being handed over to the new rich – a concern mirrored in many a local dispute over the endowments of ancient grammar schools as the commissioners' work drew attention to how their role had changed over the years.[34] Few of the trustees had specific academic leaning or experience. None had been educated at the school; very few had had sons there. They were baffled by Thring's restless spirit, his sometimes petulant, driven character and his relentless sense of purpose. In their dealings with him they reckoned him to be high-handed and unpredictable.[35] They were men of conservative outlook and financial caution, with a strong vested interest in ensuring that prudent expenditure remained the order of the day.[36] They found Thring's ambitious plans hard to understand, and they were alarmed at their actual and potential expense.[37] Having once been responsible for a school of only a few dozen day pupils, which relied for its funds only on the Archdeacon Johnson charity, they now found themselves in charge of a much larger, and much more financially complex, enterprise. However, the way in which they had allowed Uppingham's boarding side to develop meant that they now had little direct control over most of the school's income.[38]

With the chairman of the trustees living some distance from Uppingham, one board member living locally appears to have been highly influential: the figure who enjoyed much influence and prestige in this small town community, rector William Wales. The parish church and its fine rectory were right in the heart of the town.[39] Wales was a man of private means who had married well.[40] Born in Bombay in 1804, his father was the first marine surveyor-general of India, but he died when Wales was only six years old. Mrs Wales and her five sons (of whom William was the eldest) returned to England, where she petitioned the admissions committee of Christ's Hospital School to accept her son. He entered the school in 1811, and spent the next six years there.[41] Like Thring, Wales had been in Uppingham for some years by 1875, having held his living there since 1858.[42] He was now nearing retirement, and was rector supported by two curates. Among his other roles was the chairmanship of the managers of the national school in the town.[43] He was president of the subscription library, a county magistrate and president of the town's mutual improvement society. He appears to have carried out effective work as rector of Uppingham, where he and his two curates drew congregations of 500 each Sunday morning and evening.[44] He had given handsomely to the parish church restoration, rebuilding it to seat 750 people and winning social ascendancy for it in the face of growing nonconformism.[45] His information to the episcopal visitation of 1878 confirms that his was a thriving and well-organized parish.[46] Wales greatly disapproved of dissenters.[47] He had strong interests in the Society for the Promotion of Christian Knowledge (SPCK), and in educational matters generally.[48]

Even though Wales was a school trustee, it is likely that he would have disapproved of what Thring had done to Uppingham's former grammar school, by turning it into a high-fee boarding school with pupils drawn from far and wide:[49] this would hardly have been in line with Wales's own educational opportunities at Christ's Hospital. He also regretted that Thring had built a large school chapel, because it took the boys away from the parish church and established something of a rival Church of England establishment in the town – in visual terms as well as symbolic ones.[50]

Wales had also done successful work over a 27-year period in his earlier (and more fashionable) living at All Saints, Northampton.[51] He had had strong supporters there, but also vociferous enemies. There is some suggestion that he moved to Uppingham after a period of ill-health – possibly brought on by sustained opposition and controversy in Northampton.[52] His enemies drew cartoons of him and nicknamed him 'Billy Wales, the black slug'.[53] He is unlikely to have been a champion of compromise; nor was he a man with a highly developed sense of humour or subtle mediation skills. He was friendly with a number of Thring's clergymen staff (notably Hodgkinson),[54] yet he was waspish when they cited lack of time as justification for their refusal to take the extra Sunday service which he proposed to offer at the church in 1869 in an attempt to win back defectors from the dissenting chapels.[55] A number of his sermons and writings have survived, confirming the impression of an imperious, distant and aloof, but godly man with a strong sense of public duty.[56] With his keen sense of both propriety and status,[57] his disapproval would have been oppressive and keenly felt. He would not have been an easy man to deal with if crossed, or if he felt that his deeply-held principles were being challenged. His churchmanship and personality were in marked contrast to those of Thring. It would also be understandable if Thring, who had committed so much of his own fairly scant financial resources to his school, had not been at least a little envious of Wales's much greater wealth, both personal and institutional.[58]

Wales was also a leading town guardian, and the dominance of property interests can be seen when one looks at the membership of the Uppingham RSA overall. At some point during 1875 and 1877, 38 men were guardians,[59] and most of them can be traced through census and other returns. The great majority (at least 22) were farmers in and around Uppingham and its surrounding villages.[60] The landowners included many of the most regular attenders. Appendix 1 gives details of their occupations and levels of attendance at meetings.[61] The guardians reflect the fact that nineteenth-century local government was still dominated by public spirited men of good intention, but rather less technical expertise. Decision-making about the town's affairs centred round property owners, farmers, shopkeepers and small-scale professional men, many of whose occupational backgrounds suggest a town-style shopocracy with rural variations.[62] They certainly tended to be the principal ratepayers, and many were among the main employers.[63]

The union comprised 35 parishes, 19 in Rutland, 10 others across the border in Leicestershire and 6 more in nearby parts of Northamptonshire.[64] Uppingham was much the largest of its constituent communities; it covered 82 square miles with a population of over 12,000 people. The total rateable value of the parish was £9,484.2.10*d*.; the 35 parishes in the union as a whole were valued at £99,897.[65] The 25 guardians (each elected for a three-year renewable term) met each Wednesday in full session as the Poor Law union committee, but they were also members of a number of sub-committees – including one for sanitary matters. They oversaw a wide variety of local services costing between £3,000 and £4,000 annually. Their paid officials included a clerk – W. H. Brown, the same local solicitor who acted for the rector over his rents and entry fines. They also employed an overseer and collector of poor rates and taxes, an inspector of nuisances, a medical officer and public vaccinator, as well as a chaplain, workhouse master, matron and assistant, and a schoolmistress.[66] Their workhouse on the Leicester Road had been completed in 1837 at a cost of £3,128, initially for 140 inmates but later extended to 170.[67] Their sanitary sub-committee made decisions connected with the simultaneous role of the guardians as the RSA; the sub-committee's records have failed to surface thus far, but the union minute book has survived. This suggests that the guardians strove to carry out the increasing responsibilities devolved on to them since the 1872 Public Health Act carefully and conscientiously.[68]

The local government taxation returns for 1874 show that the £2,300 raised in rates and loans was already way ahead of all but a handful of RSAs in the country.[69] Of this, £1,866 was spent on sewer construction. The £2,000 loan which the RSA had taken out was one of the highest 15 or so in England and Wales by such a body; it would take many years to pay off. The amount spent on lighting and watching rates by the Uppingham guardians seems well in line with other unions.[70] Most significantly, even if one allows for the fact that figures for a single year, as opposed to those averaged over a decade, may present a distorted picture, the extent to which RSA spending in Uppingham ran well ahead of its neighbours in the year 1874, both in real terms and relative to the size of each town and RSA, is impressive.

Table 1: Spending by Uppingham and its Neighbouring RSAs: 1874[71]

Town	Uppingham	Oakham	Market Harborough	Melton Mowbray	Stamford
Total (£) contributions	2,300	304	425	452	603
Total (£) spending	2,351	172	444	275	271
Spending per inhabited house, 1871	5.36	0.28	0.88	0.27	0.19
Spending as % of rateable value, 1875	2.44	0.19	0.30	0.20	0.30

Their minute book shows that the guardians accounted scrupulously for workhouse expenditure.[72] They consulted the LGB in London, as they were legally bound to do, on a wide range of issues. They set up new sub-committees promptly in response to changing national legislation. They had put pressure on the LGB over several years up to 1875 to grant them the accelerated powers (bye-laws) which they believed they lacked. This would enable them to enforce better building regulations, to organize sanitary improvements and to borrow money or raise rates to pay for them.[73] Whether this preoccupation with bye-laws, and their desire to become an Urban Sanitary Authority (USA), was necessary is debatable; there are suggestions in the LGB papers that the Board thought that they had quite enough powers already to do what was required. Yet the wranglings over this issue took nearly five years after the passing of the Public Health Act of 1872 – until, in fact, the typhoid crisis threatened to overwhelm them. Given all the ambiguities over how the new acts should be enforced, however, this is not surprising.

Guardians such as Evans-Freke and Wales had a public duty to promote sanitary reform whilst also having a vested interest as landowners in keeping rates and costs down. This type of potential conflict of interest existed in many small towns, but in Uppingham it was made more complicated by the existence of the school of which they were trustees and the lives of whose pupils they had a responsibility to protect. As trustees they had a responsibility to set fees which were not exorbitant, yet which allowed for essential health expenditure. Moreover there was an additional dimension to any controversy about rates if the school demanded costly improvements: the extent to which, as a charity, it was exempt from some rate charges. Land which it owned fell into two categories. Portions (mostly endowed by Archdeacon Johnson in 1584) on which it had built classrooms and other educational buildings were exempt from rates – a source of further local resentment at a time when the school had moved beyond the reach of local parents. However, the boarding houses were liable for full rates because housemasters ran them as commercial ventures, and two houses appeared high up in the list of assessed properties.[74] Thring would cite the fact that 'we are large ratepayers' in justifying his sanitary demands.[75] If there were to be major sanitary expenses and rate rises at any stage, the school would not escape completely.

The chairman of the guardians was Revd Barnard Smith, rector of the neighbouring village of Glaston, two miles to the east of Uppingham. He had joined the RSA a decade earlier (1863), and had assumed the chair shortly afterwards (1866): thus his experience stretched back well before the Public Health Acts of the 1870s. His commitment to the RSA was strong and time-consuming; he was on no fewer than three of its sub-committees: those responsible for sanitary matters, nuisances (threats or dangers to property and the local environment, a sub-committee which he chaired) and education at the union school. It seems likely that busy farmers and professional men were happy to leave much of the week-to-week affairs of the union to him and, to a lesser extent, to Wales who had a much larger parish to oversee. The extent of Barnard Smith's influence can be seen in the fact that he never missed one of the 87 meetings of the guardians in a three-year period up to late December 1876. After nine years he had become weary of the burden, and had to be persuaded early in 1875 to stay for the following year.[76] The scale of his burdens, and the extent to which he must have relied on his clerk for advice on a huge range of legal, financial and other issues relating to the workhouse, sanitation and water supply, is evident in the weighty volumes of LGB papers for this period.[77]

Glaston was, admittedly, a parish of only 252 souls, but Barnard Smith ran its church and associated charities meticulously and dutifully.[78] A natural leader, his conscientiousness inspired great affection and loyalty in his supporters both in Glaston and in Uppingham. Although he had a clerical calling and status in common with Thring and Wales, whereas Thring was a classicist, Barnard Smith was primarily a mathematician. Yet his interest in mathematics was not purely academic and theoretical; as bursar at Peterhouse, Cambridge, before he moved to Rutland, he appears to have been imaginative and innovative.[79] His textbook *Arithmetic for Schools* includes problems which are inventively presented; it sold well both at home and abroad.[80] The laconic style of his textbooks suggests that he was also rather too cerebral for an impassioned and sometimes histrionic romantic such as Thring. Barnard Smith had much more in common with Wales, both in terms of interests and of temperament.[81] He had an eye for practical detail. Thring, with his big-picture mentality and flights of fancy, would have tested the patience of anyone with a tidy, economical and logical mind-set. Like Wales, Barnard Smith had no children.[82] It is significant that, unlike Thring, but in common with Wales, he was a man of considerable means.[83]

The local rating system, which Barnard Smith had to oversee and on which the union depended for its income to promote sanitary reform, had been in existence since the Poor Law Amendment Act of 1601. The fact that rates were assessed almost entirely on the ownership or occupation of land and buildings caused resistance to the system from rural landlords in times of agricultural recession when it became harder to pass payment responsibility on to the tenant.

Over the century to 1875 landowners had faced steep rises in both county and local poor rates.[84] While times had been good, this had not been a major issue – although agriculturalists had long feared that any end to price protection would put rural prosperity at risk.[85] Even after the repeal of the Corn Laws in 1846 ushered in an era of free trade, their gloomy forebodings became reality only slowly. Economic downturn came about only after a period in which farmers' profits had initially increased, thanks to a long period of good harvests and general prosperity in agriculture for nearly thirty years. In addition, greater productivity had been made possible through a combination of more intensive farming methods, better land drainage, new fertilizers and improved equipment.

At the same time the growing population boosted demand, and there was little foreign competition. New railways helped to transport produce to cities.[86]

But in the early 1870s there came a national agricultural recession.[87] The precise nature of it was complex and must not be exaggerated, but contemporary belief that it was indeed happening seems to have been real enough.[88] It was a time for landowners to restrict their spending, both personal and institutional.[89] The general gloom amongst farmers and country gentlemen had been exacerbated by the various Parochial Assessment Bills of 1860–2, which proposed a much more rigorous rating system of valuations and enforcement, and the Union Chargeability Act of 1865, which required not only that all unions provide a higher and costlier level of medical facilities in workhouses, but also that rates be assessed in the same manner throughout each union.[90] This forced wealthier parishes to assist poorer ones via a 'common fund', making such communities very alarmed at the prospect of having to bear a share of the cost of relief being given to others elsewhere. Farmers and country gentlemen feared the growing costs involved in being required to support relief (and proper public health provision) for the growing population of market towns such as Uppingham, and were concerned lest this act be a prelude to the complete centralization of all Poor Law costs and charges.[91]

The effects of a poor summer in 1873 were soon exacerbated by a very wet autumn in 1875. Together with the increased competition caused by the growing import of cheap foreign food from the vast prairie lands of the USA and Canada,[92] this recession affected arable and cereal farmers very badly, especially in the south and east of England (an area whose northern limits bordered the East Midlands). In south-eastern counties even farmers who had cut down cereal production and increased their livestock were not immune; amongst their animals there was livestock pneumonia, foot and mouth disease, sheep liver rot and swine fever. 1875 also marked the start of a period of pressure on traditional home industries as the village population drifted into towns and factories. Bishop Mandell Creighton stated that few parts of England had suffered more severely than the diocese of Peterborough.[93] This economic downturn meant declining rents and reduced returns for landlords.[94] There were demands for rent reduction, and the relationship between landlord and tenant became increasingly strained.[95] Moreover, because rents were assessed on the rental value of land and buildings, those whose main income came from rents (as opposed to profits or fees) were those on whom rates fell especially hard.[96] Uppingham had many of the former and fewer of the latter.

Some belt-tightening applied further down the social scale too. There would have been reduced spending in local shops. Earnings of Rutland agricultural workers were close to the national average for English and Welsh counties in 1867–70, and would remain so beyond the end of the century, but the county felt the effects of the 1870s depression quicker than many others.[97] Rutland was one of three counties in which small-scale rural depopulation had begun by the time of the 1881 census, and by 1911 its number of rural craftsmen would have declined by 11 per cent compared with 1851.[98] In nearby Stamford prices plummeted by 40 per cent between 1875 and 1900; working hours were cut, and the birth rate dropped – possibly in part because workers tried to limit the size of their families.[99]

In addition to all this, the prosperity of many Uppingham traders was built on mortgages in the nineteenth century, and many of these loans ran the borrower into trouble.[100] The majority of the loans were on property, and many were handed down from one generation to the next (and then added to or renegotiated, either to finance improvements or to raise capital for business ventures). Thus multiple mortgages were frequent; few were repaid in instalments, with the majority of mortgage-holders merely paying off interest every year. Many of the lenders came from outside the town, with local solicitors such as guardian William Sheild acting as go-between.[101] Some traders over-reached themselves. Uppingham's prosperity in good times was precarious, founded on borrowing which was now shown to have grown into unmanageable proportions. The shopkeepers and small businessmen feared that rate increases might be passed down to them, openly or surreptitiously, by landlords when rent reviews took place.

For all these reasons, the town authorities might well have feared a ratepayers' revolt if they launched into a bold and expensive programme of reform. The RSA's reluctance to fund sanitary reform was understandable; the town possessed few of the preconditions driving reform which affected some other communities. The population was rising but not dramatically. There had hitherto been no major epidemic of cholera or typhoid in Uppingham to throw up a campaigning figure such as William Farr, nor a popular demand for reform.[102] There had been no suggestion up to 1875 that central government might act rapidly to force improvements, and no incidence of government inspectors significantly criticizing the town's local leadership.[103] There was also no equivalent to the civic pride or financial strength which impelled sanitary and water provision improvements in some northern cities (e.g. Manchester). Nor was there much party-political activity except during election campaigns: Rutland had returned two MPs unopposed for over two decades, and for ten years its members had been Conservatives. There was no new industrial or commercial wealth to draw on (as in Birmingham). Its guardians possessed no sizeable corporate estate nor similar traditional income, such as that enjoyed by some of the older municipal corporations.[104] There was no obvious rivalry with another local town. With little or no industry, there was no economic imperative urgently to improve water supplies (as in Halifax) and no determined public health lobby.[105] Furthermore, it is unlikely that there would be huge trading profits for the guardians which they could then put towards relief of rates – a hope which drove some larger local authorities facing similarly rapid population growth to Uppingham's into promoting gas utilities in this period – especially in the north of England.[106]

There were political factors inhibiting reform too: notably a prevailing attitude amongst all classes that centralization and interventionist legislation was foreign to the national spirit. Parliament preferred specific (i.e. local bye-law) legislation to more ambitious legal enactment, and it liked to devolve a good deal of power to the localities. Its fundamental drive was still towards the elimination of abuses rather than to the creation of 'positive good'. Many of its acts were permissive rather than mandatory – even those concerning education, public health, education and industrial regulation which were passed under the Conservatives during the 1870s. The various central government inspectorates developed only slowly,[107] and where inspection did take place there was a widespread suspicion of new officials such as MOHs. Thus much depended on the initiative of each local authority; as late as 1870 only four-fifths of all Poor Law unions had built workhouses, in response to legislation now more than 35 years old.[108]

While there was a growing recognition after 1840 that action by municipalities rather than private entrepreneurs was the best way to promote sanitary reform, together with a stream of interventionist legislation designed to protect the individual (e.g. acts regulating emigration, mining and chemical pollution, vaccination, prisons and the police), sanitary reform beyond the cities came only slowly. The acts of 1848–66, which created Boards of Health, removed many nuisances (sanitary dangers) and improved local conditions, but they had little effect beyond medium-sized towns.[109] Thus in rural areas the economic, logistical and institutional demands remained formidable.

The Poor Law Amendment Act of 1834 had created groupings of parishes to be the local authorities responsible for poor relief and sanitation.[110] For administrative convenience, market towns such as Uppingham had usually become the focal point of each union, with a group of local parishes joined to them.[111] It was personally convenient for board members too: they could meet together and do business on market day. JPs were now guardians ex-officio, with the remaining members of boards being elected by means of voting papers submitted by qualified ratepayers. Larger ratepayers had additional votes.

Board members served for three years, with one-third elected annually. Not all their paid officers were efficient, and nor was the supervision of regulations, accounts and new engineering works by Poor Law Commissioners always effective. In many areas amateurism and local autonomy were still the order of the day.[112] There were nearly 700 different unions in urban and rural districts – together with an array of town councils and improvement commissions – struggling to implement reform, with little coordination between them, with tangled boundaries and rival officers levying a multiplicity of rates.[113] The separation of nuisance and sewer authorities was a particular problem.[114]

The Royal Commission of 1868–71 had recognized that guardians had done little to put nuisance removal acts into force. But now depression had brought a new difficulty: rate levels began to cause significant protests. Rate income gathered by local authorities nationally had doubled in the thirty years up to 1871 and there were calls for more of the improvement costs to be borne by central taxation – a demand not confined to the Conservative back-benches.[115] Successive governments introduced special grants as a means to lessen rate rises – but only gradually, fearing the political reaction both to increasing taxation and to increased centralization.[116] As a result, local rates rose remorselessly: in the last quarter of the century, rate revenue across the country rose by 141 per cent: by contrast, rateable values of land and property increased by 61 per cent and the population by only 37 per cent.[117] By the early 1870s the public health measures introduced in cities and towns between 1848 and 1866 to improve sanitation and water supply needed to be applied to the nation as a whole.[118] Public interest had been caught in 1871 by the typhoid outbreak in Scarborough which caused the illness of the Prince of Wales and the death of Lord Chesterfield.[119] In response to this rising concern, the *Lancet* began a special section on public health matters.

The Sanitary Commission of 1871 paved the way for Gladstone's Public Health Act a year later. This divided England and Wales into districts under specific health authorities. In rural areas this meant the consolidation of all existing health responsibilities into one local RSA – made up of the existing board of guardians of each union, backed by a medical officer, an inspector of nuisances and supporting staff. But the Commission did not lay down in detail what the duties of these authorities were, and the RSAs themselves were still reluctant to finance major reforms. It fell to Disraeli to introduce the subsequent 1875 act, which remained in force for sixty years. This set out in 343 sections a formidable list of requirements falling on boards of guardians with regard to their role as the sanitary authority on everything from nuisances, public health and infectious diseases to burials, offensive trades, food inspection and slaughterhouses. There would also now be much more onus on local authorities to ensure an adequate water supply, drainage and sewage disposal.[120]

However, stating the expectations of RSAs was one thing, but ensuring that they were speedily implemented was quite another. The habit of relying on the locality to petition for the legal powers required by its local bodies lingered on, and legislation that was merely optional continued to be the norm: as a result, any call for improvement relied overwhelmingly on enthusiasm within the locality.[121] Moreover, these new acts implied higher levels of competence to satisfy them and an increasing bureaucracy to monitor local progress,[122] so improvement was likely to be even more expensive than before. The ratepayers were still hostile to the costs: even if an RSA had the will and initiative for improvements, it would often have to spread the cost over a number of years – an issue on which it was given little guidance from London.

Poor Law civil servants encouraged guardians to think of themselves as experts, yet many of them greatly feared making a legal or technical mistake.[123] Despite their sense of public responsibility or their social aspirations, those taking on the voluntary and unpaid work of a guardian often had little or no technical expertise in the issues which they would face. Some would also have been daunted by the growing scale of the financial questions they had to deal with. Their workload and the number of meetings they had to attend had increased by virtue of their new combined Poor Law and sanitary responsibilities – which also increased their difficulties in being objective about the range of problems they had to solve. Some found it hard to prioritize. This was a period when London civil servants were also conducting a retrenchment 'crusade' ethos in Poor Law spending, so it was predictable that many guardians would decide to extend this mentality into issues of public health and sanitation.[124] They relied heavily on their paid officials, but they frequently received conflicting advice about methods, equipment and rights of way from a combination of these partial specialists and the mixture within their own membership of lawyers, surveyors and contractors. Their prolonged arguments often delayed urgent matters.[125] Even if they were able to commission improvements, there was no guarantee that these would have the desired effect.[126] Too often they hesitated to pay for outside expertise. Clergymen and doctors frequently took the lead, but they were often men of assumed or self-abrogated authority rather than of detailed knowledge. At the local level transfers of paid officials between districts were rare, thus reinforcing a cosy conservatism

There were more opportunities for RSAs to borrow money, however, and a number did so. The 1872 act encouraged them to take out loans either from private organizations or from the PWLB in London. As a result this period saw very substantial increases in loans for public works.[127] The Local Loans Act (1875) gave authorities the power to issue debentures and annuities certificates to raise loans under Board supervision. But with more loans came investigation and inspection, which showed inadequacies and defects, discovered after 1871 at least, by the new LGB's own sanitary engineers.[128] By the late 1870s the LGB had built up considerable experience in this field, and bewildered local authorities such as those in Uppingham could turn to it for advice on sewage removal, the construction of sewers, outfall works and processing plants.[129] Even then, no fewer than 27,000 different organizations and authorities of various types all dealt with separate matters in small areas and were answerable to the LGB.[130] Determined central government initiative was needed to direct them.

New sanitary laws passed at the centre of government needed to be well implemented at the local level – something which could not be taken for granted, especially in times of economic stringency. The Local Government Act of 1871 set up the LGB to exercise general supervision over local government affairs.[131] This Board combined the powers of the former Central Board of Health and the Poor Law Board, as well as some responsibilities which had previously lain with the Home Office or the Privy Council. Both town and school in Uppingham would repeatedly demand the attention of the LGB's most senior officials during the 1875–6 epidemic, but it proved ineffective in its response.

The LGB suffered from institutional problems of its own. While it was more powerful than its predecessor organizations in so far as its president was a member of the Cabinet,[132] the LGB was burdened from the outset with a huge range of responsibilities in overseeing the local execution of central government's policies – in public health, disease, prevention, housing, registration of births, marriages and deaths, and local taxation returns.[133] Yet if a uniform and efficient environmental health system was to be established, it would be essential for central government to provide expertise and experience.[134] Within months of its inception came the 1872 Public Health Act, thus further increasing the LGB's workload.[135] For a number of years it struggled to define the precise balance between setting uniform national standards and allowing (or even requiring) some very demanding local authorities to make their own decisions.[136]

There were also internal tensions. At a personal level these were between Sir John Simon and John Lambert (the Board's secretary). Simon regretted that he had to work to its senior administrators rather than direct to the president himself. He believed that his role had been downgraded, while his rival (Lambert) had the ear of successive presidents – James Stansfeld until 1874 and George Sclater Booth thereafter. Lambert respected Simon and the medical officials, but he had a firm view of their proper place in this new order of things: they should be essentially subordinate to the lay secretariat, peripheral and consultative, without powers of initiation or decision, existing mainly to provide advice when asked for it'.[137] Thus the technical expert became subordinate to the generalist administrator.[138]

The fusion of the former Poor Law Board and the medical department of the Privy Council brought together two organizations with very different traditions and expectations – which in turn led to differences over policy. The zealous employees of the medical department, headed by Simon, wished to create a cadre of central expertise in medical and sanitary engineering matters, and to force the pace of reform by *imposing* on local authorities a thorough system of central government inspection. By contrast, notwithstanding the powers given to the Board under the 1866 Sanitary Act to force local authorities to carry out sewerage and water supply works, Lambert and the former Poor Law Board personnel were happier in *persuading* local authorities to institute reform. Lambert was reluctant to cede too much power to medical expertise,[139] and Stansfeld felt that it did not need a host of medical men to improve the nation's cleanliness and purity.

Lambert and his allies were keen to win the confidence of the RSAs. Reflecting the local possessive pluralist approach,[140] they believed too in persuading local officials to employ their own professional consultants: it would reassure them that their independence was not being eroded. In addition, while the LGB would retain a number of experts in various fields for use in really contentious or difficult cases, as a general rule when it appointed its own inspectors, the key criterion was not that the appointee was an expert, but that he was a gentleman.[141] Personal influence and good judgement would count in the localities; and despite their areas of comparative ignorance they should (it was believed) be able to assess developments and the quality of local officers and those whose services the RSAs hired.

Thus the first generation of inspectors who monitored the performance of RSAs across the country for the LGB tended to come from the minor branches of landed political families or from the gentry. Many held office for decades, and often oversaw only one or two districts.[142] They continued to persuade rather than to instruct wherever possible. Significantly for Uppingham, this approach was championed by Robert Rawlinson, the Board's chief engineering inspector, who would play a part in Uppingham's events to come. He had told the Royal Sanitary Commission in 1869:

> Nothing should force me to attempt to compel a community to do what was even for their own benefit ... If persons are unwilling to receive you, you must shake the dust from your boots and go somewhere else where they will. My whole life's experience goes to this, that you cannot compel unwilling men ... you cannot put intelligence into an unwilling community.[143]

It was this reluctance for the Board to assume powers of enforcement which eventually forced Simon into retirement in 1876.

The LGB suffered from other handicaps. Its work was not especially exciting or rewarding. The scope for its officials to take decisions and to see them through was thwarted by formidable bureaucracy at the very cautious and conservative senior levels. Some of the more talented personnel found this highly frustrating.[144] Pay and prospects of promotion were poor.[145] Leadership at its very top was also mediocre in this period;[146] being LGB president was unlikely to be a job to which every rising politician aspired. Worst of all, with over 700 local authorities to supervise, the LGB was chronically overworked. It sometimes took its officials twelve months to answer a letter.[147] Not surprisingly, it tended to become bureaucratic and legalistic.[148] Overwhelmed as it was by local demands, it had insufficient time fully to provide information for central policy development.[149] It soon acquired a reputation for procrastination and inertia.[150]

Treasury control was also increasingly strict, in an attempt to keep spending under control and to minimize waste.[151] The proportion of public expenditure devoted to local government rose sharply after 1870,[152] and Poor Law retrenchment policy, dubbed by contemporaries as a 'crusade' against outdoor relief, dominated LGB spending policies until 1900.[153] New policy guidelines stated that civil servants needed to save money and so the medical department was criticized for requiring higher budgets. Requests for more or better-paid staff met with a generally negative response.[154] When the number of loan applications began to rise with the growing public awareness of sanitary matters in the 1870s, although public health reformers and local ratepayers both agreed that there was a strong case for easier and cheaper public borrowing, the Treasury continued to regard them as hidden grants. In January 1873 the Board prevailed over the Treasury with a rate of 3.5 per cent interest on loans of thirty years, but it was a short-lived victory: by 1875 the Treasury had restored a fixed-rate 5 per cent in all but the most urgent cases.[155] Under the Local Loans Act (1875), Parliament set a limit of £4m per annum on such loans. A typical PWLB loan to an RSA (depending on its size of population) was up to £10,000, and ten to twenty loans were made each month in 1876–7.[156] As a result, when seeking loans for improvements, many RSAs preferred to raise money commercially rather than borrow from LGB; they got better financial arrangements that way – and a greater degree of independence from central government.

This method of financing also made them less likely to suffer from the LGB's slow bureaucratic machine. Before it would recommend a loan to the PWLB commissioners, its own sanitary engineers had to come from London to carry out a local investigation. This often revealed other needs and problems which the local authority was then required to undertake – at more expense, and often only after extensive consultation.[157] The Public Works Loan Act of 1875 had imposed on the LGB the obligation to ensure that work was actually carried out on every project – inspecting beforehand, reviewing its progress and overseeing afterwards. This too increased its workload.[158]

The LGB was caught between the desire of both the politicians and the public for sanitary reform on the one hand, and their distrust of increased government centralization and its cost on the other. In these early years of its existence it also had to give advice on a huge range of minor matters to RSAs still uncertain about their precise responsibilities and powers of decision-making. The writers of *An Outline of Local Government and Local Taxation in England and Wales*, written in 1884, concluded: 'The defectiveness of local government overwhelms the LGB'.[159] The LGB's experience of Uppingham's minutiae, and Uppingham's experience of its delays, suggest that this was indeed the case. It was likely that any significant crisis would overwhelm both of them. Given the range and scale of problems which confronted both the LGB and the Uppingham guardians – vested interest against rate rises, agricultural depression, suspicion of official-dom, limited capability and resources of expertise, and stifling bureaucracy – it was highly likely that one or both of them would be overwhelmed by the crisis about to break over their heads.

3 LOCAL MEDICINE AND LOCAL DOCTORS

In 1865 the medical officer of Seacroft, West Yorkshire, painted a graphic and alarming picture of his local area:

> None of the villages have any drains whatever. It is the practice to throw everything in the shape of sewage, garbage, refuse, and even solid excrement into the highway, on to the green and the adjacent midden [waste] heaps. And into a ditch if such be handy ... Almost every cottage has in front of it a midden-heap ...[1]

It is likely that Uppingham a decade later was very similar. However the authorities in the capital might regret it, and whatever the advances in medical science being discovered there, sanitary provision in rural England had changed comparatively little by the late nineteenth century:

> Britain had managed its sanitary arrangements quite well for a number of centuries, drawing water from relatively unpolluted sources, disposing of waste without difficulty. The village drew water from wells and streams, distributed excreta over the fields. Prolonged drought caused a shortage of water and a stink. 'Fever', never entirely absent from the rural community, erupted into local epidemics. Infant mortality would rise and the medical man [would] talk wisely of 'summer diarrhoea' or 'infantile cholera'. When the rain came, as it always did, the water level of surface wells rose quickly, dried ordure leached into the soil, and the village reverted to its normal condition of too wet rather than too dry. It had worked for centuries without producing unbearable conditions and persisted in remoter country districts until almost the present day ...[2]

The nineteenth-century population increase put an unprecedented strain on these arrangements. The population of England and Wales had virtually doubled in the five decades to 1850. In many cities it was densely packed into low-quality, low-cost housing with few planning controls, thus making disease an increasing threat. After that date water closets gained in awareness and popularity, partly because the organizers of the Great Exhibition in London (1851) had provided public lavatories for those who visited it. London generally had water closets by 1870,[3] although their respective merits versus dry earth systems would be disputed for some years. But London represented a very different world from Uppingham.

In rural areas there were fewer pressures from industrialization, but health hazards were just as real. House sanitary arrangements were primitive; people often used holes in the ground at the back of the cottage, or relieved themselves in alleyways or in fields with streams. Livestock grazed and wandered at will, leaving the inevitable evidence of their presence. Uppingham had seen a large number of the back yards to its properties filled in with new dwellings during the nineteenth century; parts of the town had been densely rebuilt, and many people lived in the buildings in which they worked – a hazardous situation in health terms.[4] Sir John Simon had drawn attention to such dangers in his report to the Privy Council a year earlier (1864), and had complained that local authorities were doing little or nothing to remove them.[5] House waste and excrement were often placed in simple pits, into which were thrown additions of soil or ash to cover the contents, turning it into a more or less solid mass. Residents often threw in slops or other domestic refuse too. Gradually large buckets or closets were introduced which held sizeable quantities of sewage. Originally pervious, newer types of closet were developed through which waste material would not leak. Medical authorities called repeatedly for pail closets to be created inside houses, using either ash or water. By the end of the century there would be clear evidence that pails (as opposed to privies) reduced the incidence of typhoid,[6] but even where RSAs made it a high priority to speed up improvement, bye-laws to force residents to conform were often inadequate.

Where cesspits existed, significant amounts of water were needed to drive the waste from house or privy to pit. This was hard to achieve because, with no piped water provided, it had to be manually drawn from wells. Drainage gradients needed to be generous if the pits were to be efficient and to prevent the build-up or rushing-back of waste when systems were full, especially where drains were shared between houses. They needed to be free from leakage – and sited a careful distance from the house if cellars were not to become flooded. Cesspits, like the privies and closets, needed careful and regular emptying by the scavenger or night-soil man to well-chosen sites – if the householder did not bury it himself in the garden. Pits were often left to overflow until the local farmer was ready to collect the waste. Above all, drains from houses, cesspits which might leak and sewers across towns and cities needed to avoid watercourses and wells if river pollution and cross-contamination were to be avoided.[7]

Such problems were resolved only slowly. Sewer commissions were ancient institutions, but they existed mainly to deal with surface water, and there was no legislation before 1835 to compel house drainage into public sewers, even in cities. By 1870 cesspools were disappearing in towns, but in the countryside many still failed to treat sewage before it disappeared into the ground, often near watercourses.[8] Animals continue to graze undisturbed.[9] The cost of regular checking and cleaning of cesspits by the local authority was often deemed prohibitive. Ultimately, every system needed somewhere for the waste to be stored or disposed of. There was too much scope for spillage, and for waste to be left on floors and in corners as it was dug out. Solid waste systems needed emptying manually – usually to large mounds on the edge of the town or village. These sewage piles were sometimes carelessly sited – too near to wells and springs, or in places where humans would tend to pick them over and animals to graze or roam on them.

Uppingham appears to have been no worse than its neighbour towns in respect of its sanitation, and possibly better.[10] The East Midlands was not in the forefront of sanitary reform. The city of Leicester (only 20 miles away) had been described in mid-century as being 'honey-combed with filthy, stinking cesspits and middens, polluting air, soil and water'.[11] It headed the urban league table of annual infant mortality rates between 1860 and 1899.[12] With only two official visitors – compared with 25 in Liverpool – after the Notification of Births Act was passed in 1907, it also had one of the highest infant incidences of diarrhoea in the country.[13] It had relied until 1850 on cesspits, and even as late as 1875 (during a time of rapid population growth)[14] only parts of a sewerage system were in place there: much of the waste still discharged into the river. A pail closet and emptying system had recently been started, but privies were still widespread – a system notoriously prone to typhoid, according to contemporary reports from MOHs in Nottingham and other cities.[15]

The comparison with another neighbour, the county town of Oakham six miles to the north of Uppingham, is instructive. There had been bitter complaints from Oakham residents in 1856 that drains could not cope with demand from the new water closets, and that those using the latter would have to revert to cesspool drainage. Unfavourable comparisons were made with Uppingham, which was said to have spent many thousands of pounds on improvements.[16] Twelve years later a civil engineer reported that he found the sewers in Oakham to be:

> Generally of a most primitive construction [with] rubble stone side walls with slab bottoms and covers ... neither the material of construction nor of subsoil can be water tight, and from their superficial nature [they] must always be liable to pollute the water in the surrounding wells.

He recommended larger diameter pipes at greater depth but, possibly fearing that an expensive scheme would be rejected out of hand, his suggestions were modest: the estimated cost was only £600 – significantly less than Uppingham had recently spent.[17] Unsurprisingly, three years later (1871) the local paper again reported 'an abominable stench' near the marketplace as cholera loomed in the area, resulting in a slightly more ambitious second scheme (£700); deep

sewerage had to wait until 1878 when the town once again followed in Uppingham's wake.[18]

Water provision was not much better. In the early nineteenth century there
were few municipal water facilities, and a growing problem. Liverpool had supplies barely sufficient for cooking and drinking. London and Bristol fared better;
they had more wells, but these were very liable to cesspool soakage and, as water-
closet usage grew, an increased risk of cross-infection.[19] Urban expansion led to
increased demand, and simultaneously to the destruction or pollution of many
traditional water sources as slaughterhouses and bleach-works appeared on river-
banks and building activity obliterated or infected springs. Cholera and typhoid
became more frequent. There were increased demands for regular washing and
bathing – and for better water supplies.

As technological developments made for better and larger reservoirs and the
filtering of water, cities varied in their level of uptake. By 1879, 79.4 per cent
of Manchester houses had an internal supply, in contrast to Birmingham where
there were 43,000 houses without one as late as 1913.[20] Piping did not solve
all problems: the sporadic water supply, its cost to consumers and the limited
sewer networks restricted the rate at which water closets spread into upper and
middle-class households.[21] Even though plumbing became more professional,[22]
there were still many sources of impurity: some houses stored water in wooden
butts which had been wine or beer casks. Some butts were poorly covered, or
stood in warm kitchens. Until cast-iron pipes were introduced to replace those
made out of wood and lead, pipe joints tended to be suspect.

The poor fared worst. Many homes in the years up to 1900 still relied on
street standpipes,[23] with closets flushed through the use of the hand-pail, giving
rise to serious bacteriological and health risks. Not until 1914 did most urban
areas have piped supplies, by which time politicians and others had become
convinced of need for municipal management and were looking to undertake
ambitious schemes from further afield. By then, Manchester, Liverpool and
Glasgow had all begun to pipe in water from natural sources well beyond their
confines, but regional cooperation between suppliers was rare.[24] Most towns
depended upon a combination of common watering-places, private and public
wells, ponds, reservoirs, streams, rivers and springs, and rain water – with local
government exercising overall responsibility for safeguarding supplies.[25]

Until the 1800s most man-made water provision outside London was carried out by town (i.e. public) authorities. The nineteenth century saw a gradual
shift towards private schemes before the final years brought the municipalities
back into the field, sometimes expensively buying out the established companies.[26] Parliament tried to prevent both monopoly abuse and inter-company
rivalries, but it recognized that while entrepreneurs (who included some MPs[27])
tended to act for short-term gain, they could get things done quickly at a time

when borrowing facilities for municipal authorities were very limited. However, the doubts which grew about the inadequacy of private waterworks resulted in a resurgence of municipal activity because it was believed that 'water was the key to the public health problem'.[28] This led to legislation in the 1840s: if the company agreed, town authorities were permitted to assume control over private undertakings,[29] although the cost of private bills acted as a deterrent. By 1871 only a third of all the urban sanitary authorities owned, or were developing, their own water supply – either themselves or through privately owned third parties. Only with the Public Health (Water) Act of 1878 was an easier legislative path provided for the municipal purchase of private waterworks.[30]

Rural communities were expensive to supply with water, and had limited funds with which to pay for installation.[31] The 1872 Public Health Act obliged both urban and rural sanitary authorities to provide a supply, but in many cases it was a very rudimentary one. Many homes had only one tap[32] and the less fortunate had to rely on street standpipes. Many had water for only a few hours and only on some days per week.[33] Some authorities prevaricated on grounds of expense or ignored the requirement of the Act altogether. As a result, the 1875 Act went further, requiring the guardians to acquire the private companies if they were proving inadequate to their task, and putting in place an arbitration procedure for disputes.[34] Three years later the 1878 Public Health (Water) Act stipulated that ratepayers must pay for proper sewage farms. But RSAs frequently cut costs on improvements.[35] Only by the eve of the Great War had the municipalization movement transformed the water industry into one of the most collectivized sectors of the British economy.[36]

Meanwhile there was increasing analysis of water impurities. In an age before bacteriology, those inspections which were carried out tended to concentrate on the water's visual state, or its chemical additives. Murky water polluted by chemicals, minerals, sewage or other rotting material was easy to spot; thus clear, sparkling water was frequently taken to be a sign of purity – yet it might easily hide just the pathogenic organisms which caused cholera, typhoid or other diseases.[37] The later nineteenth century gave rise to a number of experts (some real, some self-appointed) and plenty of controversy, but Christopher Hamlin concludes that:

> Good water analysts at the end of the century were probably about as good at detecting dangerous contamination as were good water analysts fifty years earlier, and views on which waters were good and which were bad changed relatively little ...[38]
>
> The coming of bacteriology in the 1880s only transformed the idiom of the debate, without resolving the key issues: whether water could be purified reliably enough to be safely used by the public and whether any means of water analysis could reliably demonstrate that purification[39]

What seems evident is that there were many more medical people taking an interest in it by the 1870s, they were being held more rigorously to account for their findings and they were no longer making extreme claims about the reliability of those findings.

Outside the major cities and towns, in water supply as in drainage the progress was slow. Again the East Midlands was no leader: Leicester had no piped water at all until the 1850s.[40] In Stamford, a medium sized town only eleven miles from Uppingham and significantly larger, there were severe outbreaks of typhoid in 1868 and 1869. A year later, a report bemoaned the fact that the town's underlying geology had been broken up by building, quarrying and natural forces. Only a few of the streets possessed sewers, and the river (as it passed through the town) was 'a most offensive cesspool' and still liable to frequent flooding. The Marquess of Exeter at nearby Burleigh House supplied water to parts of the town by an Act of Parliament of 1837; but other parts still relied on fifteen pumps scattered around the town's streets, and some would remain un-piped ten years later. The council spent seven years between 1870 and 1877 debating how to improve things, and in the following year its MOH condemned the existing supply's impurities and offensive smell. Only in 1882 were improved arrangements made, and one of the original conduits remained in use until around 1900.[41]

In Oakham in 1868 there were 'hundreds of poor families who have to go two miles for fresh water'[42] and when deep sewerage was installed in 1878 many of the springs were diverted. This resulted in an acute water shortage at the east end of the town, which still relied on 252 wells and three public pumps in 1897. Pollution from the sewerage system got progressively worse. Although the town acquired a flushing tank in 1880 and its first piped water supply in 1885, the latter was very small in scale and a decade later many wells were condemned as suitable only for washing. Attempts in the 1890s to find a new source foundered because the water proved to be so hard and the cost so high; it was 1900 before the Oakham water company came into being and completed the provision of an adequate supply.[43] Thus Uppingham was far from alone in relying on wells. The fact that it had neither a publicly nor a privately funded system of piped water is also unexceptional.

Meanwhile, faced with such health hazards, country doctors were in the front line of the battle against disease.[44] They fought to alleviate the effects of colds, coughs and influenza in winter (often leading to bronchitis or pneumonia), and with diarrhoea, typhoid and other fevers in the summer and autumn months.[45] Fevers, chest and throat infections were especially prevalent compared with modern times.[46] The term 'general practitioner' (GP) was introduced in the 1820s to denote those who practised all types of medicine, including surgery, midwifery and pharmacy. Such men came from a wide variety of backgrounds, and the social divide between the university-trained physician and the 'apothecary-doctor' (which had once been absolute) had narrowed. A doctor had several possible routes through training: he might become a Licentiate of the Colleges of Physicians and Surgeons (LRCP), or of the Apothecaries' Societies (LSA), or of one of the universities. The LRCP of London was the qualification required by the LGB for any doctor practising the complete range of medicine, surgery or midwifery. Some GPs took qualifications in only one of these areas; a small number took higher qualifications such as Member of the Royal College of Surgeons (MRCS) or Fellow of the Royal College of Physicians of London (FRCP), or the Fellowship of the Royal College of Surgeons (FRCS).[47]

The Medical Act of 1858 established state registration of qualified doctors, and set up the General Medical Council to govern the profession. GPs could register on the basis of a single qualification in either medicine or surgery; the former system of medical training based on individual apprenticeships or pupillage followed by 'walking the wards' of a hospital had already given way to more systematic training. But GPs' training and status did not necessarily imply a high degree of expert knowledge. New medical discoveries were not quickly handed down from laboratory scientists in Europe's capitals to rural GPs. The medical schools were too much geared to the production of specialists and not enough to the needs of the would-be GP. Only later would the concept of general practice as a specialism in its own right be recognized. Meanwhile there was a strong emphasis in medical schools on academic cramming but little on training in the practicalities needed by the generalist.[48]

Doctors were taught to diagnose symptoms as best they could. In 1875, when typhoid broke out in Uppingham, William Farr considered that 'diagnosis, though still imperfect, has within 35 years made remarkable progress'.[49] Even so, both diagnosis and prognosis would remain inexact skills for some time to come. While GPs dealt humanely with their patients, there was little emphasis in their training on precise measurement, and few effective curative drugs were available. Often there was little that a GP could do except to 'reassure the patient and console the relatives'.[50] With the passing of the 1874 Births and Deaths Registration Act there was an increased expectation on doctors to certify and notify cause of death. The Infectious Diseases (Notification) Act 1889 would require the metropolitan local authorities, and permit provincial ones adopting the Act, to demand prompt notification to the MOH of all cases of smallpox, asiatic cholera and many fevers.

Medical men had traditionally enjoyed high status and social respect. One mid-nineteenth century writer on rural life considered the apothecary:

> one of the most important personages in a small country town ... He takes rank next to the rector and the attorney, and before the curate; and could be much less easily dispensed with than either of these worthies.[51]

But by 1875 country GPs were fighting to establish themselves in a social hierarchy with a large number of newly-qualified legal and medical practitioners.[52] Within the profession there was an instinctive mistrust of going into partnerships. Specialists represented a growing threat to the livelihood of the GP, who needed a core of middle-class, fee-paying patients to offset bad debts amongst poorer patients. Not only that, but GPs were battling with each other for patients, and to establish a niche market in each town, as increasing numbers of newly-qualified doctors emerged from medical schools.[53] There was also a variety of practising unqualified assistants and counter-prescribing chemists, as well as homeopaths and a number of quack doctors and other charlatans.[54] When children became ill, many parents relied as far as possible on home remedies to effect a cure, with herbs widely used. As a last resort the Poor Law medical officer might be called in.[55]

The over-supply of medical services in relation to market demand was thus a widespread problem for the GP.[56] Livelihoods and incomes had to be built up carefully, and nurtured over a period of years. Younger sons of medical families often followed in their father's footsteps, and there were many multi-generational family practices, thus strengthening a sense of territory. GPs needed organizational and entrepreneurial skills; many worked from a room at home, with their wives acting as bookkeepers and practice organizers. Rural GPs made many more home visits relative to surgery contacts than their counterparts in towns, and travelled greater (and costlier) distances.[57] Those who were able to augment their income by appointments as public vaccinators, coroners, medical witnesses, workhouse medical officers, or agents of births and deaths registration and infectious disease notification, fought hard to keep their positions. So did those appointed as medical officers to schools or town sanitary authorities, or even as MOHs.[58] Faced by all these demands and pressures, it is small wonder that some doctors fought tenaciously to preserve their territory and reputation.

Uppingham School had a sanatorium on the Stockerston Road, built in 1869 by the masters for £3,000.[59] There was no town hospital as such, although indoor relief had begun in 1734,[60] and the union workhouse had been envisaged as a multi-purpose building: part workhouse, part orphanage, part old peoples' home and part unemployment centre. Medical books and patent medicines were widely available in local shops.[61] By 1875 three doctors served the town and its surrounding community.[62] This level of provision, in a town of 3,000 people in term-time and fewer in the holidays, does not suggest a shortage of medical expertise, given that the ratio of medical men per head of national population at the time was just under one to 1,700.[63] The competition for custom may well have been intense between the three doctors: Augustus Walford (aged 54), Frederick Brown (aged 40, and brother of W. H. Brown, the RSA clerk), and Thomas Bell (aged 39, and the most recently trained). Bell held both LRCP and

MRCS qualifications dating from 1861; Walford and Brown were only qualified in surgery and as apothecaries (MRCS and LSA) – qualifications gained in various years between 1854 and 1872. In addition to his medical practice, Walford was responsible for medical matters at the workhouse and was the town's public vaccinator.[64]

Dr Bell's family had long-standing medical roots in Uppingham. His grandfather (James) had been an explorer who 'graduated at Edinburgh in the 1770s and settled in Uppingham in 1780 after a voyage to the polar regions as a surgeon on board a whaler'.[65] James became an apothecary and a pillar of the local congregational church, where five of his children were baptized. One of them (John, father of Thomas) had practised medicine in Uppingham for many years, and was now in his eighties, living in retirement in the town.[66] Thomas Bell ('Thos') was not only a GP; he was also the school's medical officer (MO). He lived in High Street West with his wife and four children aged between four and nine, right alongside the school boarding houses that would become typhoid stricken.[67] He would have felt all the economic and territorial pressures which faced GPs, even though he had lived in the town all his life and had worked in partnership with his father in the years after he qualified. He had an emotional as well as a professional attachment to the school: he was the fifth of seven brothers who all passed through it, and he had entered it as a day-boy in Dr Holden's time (1846) when he was scarcely nine years old. Apart from a few years in London at medical school, he would spend the rest of his life living within a few yards of the school.[68] His obituary article in the *USM* in 1914 states that after qualifying in London in 1861, he had returned to Uppingham with good references from his tutors. There were those who praised his support for Thring through his 'calm judgement (and) unfailing care'. It is likely that he was a shy man:

> Absolutely devoid of anything like self-advertisement that he hardly did himself justice ... his reserve was, in great part, repression – an element in the deliberate sobriety of his judgement. It was often, but mistakenly, thought to be indifference. A more buoyant manner, a more enterprising temperament might have made his work brighter and more attractive; but one may doubt whether in that case it would have contributed so much to the building up of Uppingham ...

Bell was clearly conscientious. He had outside pursuits – notably a keen interest in natural history: 'knowing every inch of the countryside around for miles'. But he lived for his work:

> Only those who knew him well knew how he took pains to keep himself abreast of medical knowledge, and how often [a] great part of his rare and short holidays was spent in visiting some large hospital ... His work with his poorer patients was splendid, and in their cottages perhaps he was seen at his best. He was unremitting in his patient care of them, and would himself give personal attention to petty details of

nursing ... He was a conspicuous instance of the kind-hearted helpfulness so common among medical men.[69]

Reading between the lines of this generous and affectionate tribute, however, it is clear that in medical terms he was not a high-flyer. It is likely that his achievements came by hard graft rather than brilliant, rapid diagnosis: 'By high aim and steady persistency he achieved results as good, may be, as those attained more quickly by others'. The writer added: 'Though housemasters may at times have longed for a quicker decision in a suspicious case, they seldom had reason to regret his deliberation'.

There was another, new pressure on GPs such as Bell: the growing supervisory power of the MOH, especially in times of medical difficulty. The previous half-century had seen a marked rise in public interest in health matters, to the point at which it had become a moral crusade. Echoing John Wesley's earlier assertion that 'Cleanliness is indeed next to godliness',[70] the *Servants' Magazine* had defined the issue in 1839: 'Dirt is the natural emblem and consequence of vice ... Cleanliness is an essential in a servant ... a dirty Christian is considered all but an impossibility, and a contradiction in terms ...'.[71] The reduction or total elimination of dirt in the form of poor drainage, foul water, manure heaps and other public nuisances became a cause taken up by sections of the medical profession – and by the medical press. The *BMJ* became the public health mouthpiece,[72] and in a number of articles in 1876 it reviewed the state of the nation's health. The death-rate from fever 'including typhus, enteric and simple continued fevers was also lower than for any year for which records exist'.[73] Even so, there were still 12,500 deaths per year. It concluded that sanitary authorities must be even more active:[74]

> The Registrar-General's quarterly bulletin ... possesses an interest which is constantly increasing ... Death-rates and their details have become almost popular topics. Public health is already feeling the change in public opinion ... There is satisfaction to be derived from the evidence of the decline in urban death-rates, in spite of the increasing density of our town populations. That rural death-rates do not show a similar improvement can only be accepted as evidence that in health matters there is no such thing as standing still.[75]

One response had been the creation of a network of MOHs – part of the wider growth of new local government positions which included sanitary engineers, food and drugs specialists, building and factory inspectors and town clerks.[76] This replicated the earlier developments in cities after Chadwick's contribution to the Poor Law Commission in the 1840s. Liverpool (the city from which the largest number of doctors' sons came to Uppingham) appointed William Henry Duncan as its first MOH in 1847. The Association for Medical Officers was founded in 1856. As MOH for the City of London (1848–55) and then working for central government as MOH to the Privy Council (1855–76), Sir John Simon created a medical department to oversee public health and factory legislation, and to promote scientific research and vaccination. Simon planned for this department to supervise the medical profession as a whole, believing in the need for a strong central authority (as Chapter 2 showed), with one single government ministry presiding over both the Poor Law arrangements and public health. A registrar-general would monitor disease and sickness statistics. Local authorities, each with an MOH, would call on the expertise of a medical and engineering inspectorate when they needed it. The government accepted all these ideas, and incorporated them into the 1872 Public Health Act. Simon himself became Chief Medical Officer of the new department which the Act created – the LGB.[77]

Thereafter, most major cities appointed their own MOHs – albeit reluctantly in many cases.[78] Some cities shared local ratepayers' wariness of both the increased expense and the implied centralization of government control. In Birmingham, 'only when the Public Health Act of 1872 made the appointment of a medical officer compulsory for even the smallest rural district did the corporation of the third biggest English borough reluctantly step into line'.[79] In Leeds, the Corporation kept the salary of their MOH unchanged for six years; he promptly left and they advertised for his successor at an even lower salary of only £300 per year – with the result that they attracted a very low-calibre successor whose incompetence and neglect contributed to the Headingley typhoid outbreak of 1889.[80]

However others, including Robert Weale, the national Poor Law Inspector, believed that in rural areas the local Poor Law MOs and other staff could easily undertake sanitary functions, at significantly lower cost. These arrangements were seized on with enthusiasm by ratepayers, who (with most public health legislation being permissive rather than statutory) were happy to follow the retrenchment thinking of the Poor Law officials during the 'crusade' period, and to contract doctors to combine District Medical Officer for the Poor (DMOP) and MOH work – and simultaneously to reduce the number of DMOP posts within individual unions.[81] The prospect of poorly-paid, hard-pressed and technically untrained doctors becoming MOHs in innumerable small local districts filled Simon with dismay – even though they had to send regular reports to the LGB.[82]

With so much inertia and opposition there were still only 50 or so MOHs by 1872.[83] The 1870s Public Health Acts, however, made MOHs into prominent figures in their own domains and forced the rural districts to make their first appointments. They had to be qualified medical practitioners, but their skill and knowledge was variable – as was their determination and zeal. Some were almost messianic in challenging local employers and councils to put money into improvements; others were content to settle for a quieter and less contentious life. There was little attempt by the LGB to specify the type of person to be appointed, or the level of expertise expected from them; the LGB could provide expert medical back-up where necessary. It was not thought necessary to debar them from seeing their own individual patients.[84] Many authorities had two big problems in finding suitable candidates. The work appealed only to a certain type of doctor. Even if he remained a medical man (which most strove to do) rather than becoming a bureaucrat, after the first generation of able and imaginative MOHs left the scene, high-quality successors were harder to recruit.[85] In many rural areas – including Rutland – the population was so small that they could not afford a full-time appointment or an officer of high calibre without a huge increase in the rates – despite some subsidy.[86]

Simon had been appointed in London in the 1840s on a salary of only £500 p.a. – later raised to £800.[87] Large towns paid up to £1,000 by 1900, but many MOHs still earned only £200–500;[88] a mere 6 per cent of them could afford to be full time and many were on short tenure. Some districts dragged their feet: fewer than half the rural authorities had filled their vacancy by 1874.[89] Some appointed men of low calibre. Other RSAs joined forces to pay a salary to attract a stronger candidate. This initiative created MOH-led districts – yet another set of territorial units within local government. The *BMJ* protested in an editorial in May 1874 entitled 'Fancy-work in Sanitary Organization':

> There is something manifestly absurd in officers constituting their own authorities, yet this is in effect what these gentlemen propose to accomplish ... Would it not be wiser to call on Parliament ... to [make their boundaries coincide] with the older and more successful administrative divisions of the kingdom ...?[90]

This first generation of MOHs did not have an easy time. Their powers were poorly defined by the 1872 Act, whose authors had envisaged them largely as analysers and record-keepers rather than leaders in a new field. They campaigned for stricter disease notification.[91] Yet they had a huge range of new legislation within their remit if they saw fit – and many new ideas to bring together if public health and preventive medicine were to be established as a priority in the public consciousness. Local GPs resisted the MOH's intervention in diagnosis of their patients as a threat to their authority, and from 1889, when the Infectious Diseases Notification Act was passed, to what they saw as its onerous procedures.[92]

Householders who welcomed their local doctor often resented the MOH – especially if he recommended the removal or isolation of a patient.[93] There was a stigma about infectious disease. The ratepayers resented the expense; within months of the passing of the 1875 Act, a meeting of MOHs for some of the new combined districts throughout England was calling for stronger tenure arrangements against local RSAs and angry ratepayers calling for their dismissal.[94] In

Lincoln, when the corporation considered establishing a local board of health (1866), the 'economists' sent threatening letters to its members, one of whom was told that his coffin had been ordered; another resigned after a similar death threat.[95] Local councils could be miserly with expenses as well as salary. In larger areas travelling could be extensive, and MOHs had to use their own gig, horse and groom.[96] Many were sustained only by a passionate belief in the value of their work – 'a new cadre of officials who devoted their lives to improving the lot of the urban poor ... it was [they] who actively enquired into reasons for stubbornly high levels of mortality'.[97]

Given all this, Bell and Thring might well have found in 1875 that there was still no MOH serving Uppingham at all, or a low-level officer promoted from being inspector of nuisances, perhaps someone timid and out of his depth.[98] As it was, they found themselves dealing with the MOH for one of these new combined districts. The Uppingham RSA had joined forces with a number of others in the Northampton area, and had appointed as their MOH a medical practitioner of substance, intelligence and iron determination: Dr Alfred Haviland.

Like Thring, Haviland came from a Somerset family. Both his great-uncle and father were surgeons in Bridgwater; his father's first cousin was John Haviland, at one time Cambridge University Regius Professor of Physic and Fellow of St John's College.[99] Born in 1825, the fifth of eight children, Alfred Haviland qualified from University College Hospital, London, in 1845 and became (like Dr Bell, a near-contemporary in age) a partner in practice with his father. In 1849 Bridgwater suffered a major cholera outbreak lasting four months, with over 1,000 cases and 200 deaths, believed to have been caused by water taken from wells which had been cross-infected from fields of rotten potatoes.[100] This gave Haviland first-hand experience of the dangers of epidemics – and of local demands for a better water supply.[101] He was appointed to practise surgery at Bridgwater Hospital, but his career was cruelly cut short when he poisoned his finger during an operation in 1867 and nearly lost his life.[102]

Haviland's school of medical opinion set particular store on physicians having a thorough knowledge of local climate, geology and natural history, as well as living conditions.[103] During the cholera outbreak, he took meteorological observations day and night. He established that cholera cases increased after 'calms' of weather, reducing as soon as the south-west wind returned. In 1855 he produced his first major book, *Climate, Weather and Disease* – used as the basis for part of the Registrar General's Annual Report nine years later. He also tabulated ten years of death rates from these reports (1851–60) to demonstrate the geographical distribution of heart disease. So began a lifetime's interest in medical mapping. Haviland took this skill to previously unknown levels of sophistication. His work included studies of cancer (1868), and dropsy and phthisis (1875). He discoursed on clays and limestones, on medical geography, on the merits and shortcomings of Brighton, Scarborough and the Isle of Man as health resorts, and on the follies of hurrying to catch trains – a mode of transport which, it had been suggested, tended to cause heart attacks – in an essay entitled 'Hurried to Death'. He became an honorary lecturer at St Thomas's Hospital, and was awarded the silver medal of the Royal Society of Arts in 1879 – which caused the *Lancet* to comment: 'His maps are unique, and have rendered a great service to medical science'.[104]

Haviland was a tireless writer and lecturer, mostly on topics related to climate and disease. He wrote forcefully, as in his report on the Kingsthorpe area of Northampton, which he described as having:

> ... a magnificent supply of pure spring water, [which] is in a most loathsome condition. The water is contaminated with the filthy oozings and drainings from slaughter-houses, wells converted into cess-pools, obstructed drains, muck heaps and surface water ...[105]

He frequently courted controversy; indeed in many ways he seems to have thrived on it, even becoming involved in a fierce dispute with his relatives over the family genealogy.[106] His lectures, books and papers were summarized and reviewed in both the *Lancet* and the *BMJ* on a number of occasions between 1872 and his death in 1903 – mostly but not always favourably. An unnamed reviewer in the journal *Athenaeum* in 1876 praised the quality of Haviland's maps, but he also identified major problems with his methods and results, and believed that his opinions were hasty and arbitrary.[107] He further suggested that Haviland had little idea about the true cause of typhoid, being a 'closet miasmatist [whose] idea of air-sewage was basically that of a poisonous gas or noxious fumes'.[108] Modern writers have contrasting views: T. W. Freeman, in an article on Haviland's medical mapping, describes him as 'a devoted worker', but Frank A. Barrett believes that 'although his technique was innovative, his analysis was flawed'.[109] Barrett concludes that Haviland was so determined to demonstrate the value of medical mapping that 'a grave weakness of [his] analytical ability was that when he encountered contradictions he did not believe that these facts pointed to errors in his conceptualization of the aetiology'.[110]

Haviland was appointed as the first MOH for the scattered Northampton districts (spread over four counties) in April 1873.[111] His salary was unusually high by the standards of the time, at £800 per year.[112] The *BMJ* published regular lists of new MOH appointments in this period, mostly at less than half this sum, and it records that there were 63 candidates for the post.[113] Within a year Haviland was involved in a spirited exchange of views with his new employers over their refusal to pay the costs of publishing his first, lengthy annual report – which he then produced himself as an offprint from the local newspaper.[114] He had a vast territory to cover, cutting across all sorts of existing sanitary, Poor Law and local government boundaries. Dr Henry Rumsey, president of the Gloucestershire Branch of the British Medical Association, used him as an example of all that was wrong with the new system: 'poor Mr Haviland having to travel over a most extensive and impracticable circuit imperfectly supplied with railways'.[115] Its boundaries were hardly logical, as it included parts of three other counties, but without the Brixworth and Kettering unions which lay geographically in its centre.[116]

Haviland set to work with a will. His first annual report on Northamptonshire is a comprehensive document, descriptive and statistical, which drew heavily on his mapping methods.[117] Considering the short time that he had been in post and the geographical extent of his area, he had worked fast. It seems that he had a good eye for detail, although the suspicion lingers that some of it was random and designed merely to impress his employers.[118] He concentrated particularly on recent issues relating to fever, believing that typhoid was 'a national disgrace; we ought not to rest until we reduce it to one simply local or personal; its existence will then become punishable'.[119] He observed that it was generally contracted either through infected water or sewer gases or by contagion, but significantly he declined to commit himself in the miasma/contagion debate: '[It] is very-present amongst us but is perhaps one of the most easily preventable of the many forms of death with which we now have to deal', and 'I believe that the disease and death returns of typhoid fever are the best indicators that we have of the sanitary condition of any place, whether in the town or in the country'. He praised the impact of recent legislation, but he stated that further progress could only be made if the powers of the RSAs were strengthened. He rounded off:

> I have made typhoid the subject of my annual report simply because I believe that it is the disease which teaches us best how to be prepared for others. Let us once successfully subdue this great but preventable cause of death, and I feel assured that we shall be able to cope with its allies. Let us strive to get pure water, cart away our unpolluted sewage, and institute a system of scavenging throughout our villages ...

A strong supporter of ash closets, he attacked those who believed in water-based sewage systems: 'I believe [then] we shall hear less and less of sewage farms, until the time shall arrive when they will become things of the past, like their progenitors the water closet'.[120]

In Uppingham itself in 1874, Haviland found an authority which saw little need for urgency. Compared with Leicester its problems were (he judged) small; its population was less dense, its incidence of most diseases was lower and, having no railway, it was comparatively vagrant-free.[121] The recent level of activity of its RSA – better than its neighbours, but with much still to be done – seemed justified by the annual reports of Simon's department, which suggested that its state of health was no worse than countless others. Haviland concurred with

Simon: he stated that, compared with neighbouring authorities, Uppingham's state of health in 1873 was good. In England and Wales as a whole, the figure for deaths in the years 1861–70 had been 2,242 per 100,000; in the Uppingham registration district it was only 1,846 – lower than all the other fifteen districts in the Northampton combined districts.[122] Deaths from fever, diarrhoea and diphtheria in Uppingham had all been low in the 1850 and 1860s – as had the number of deaths of those aged under 1 and under 5 years old. Only scarlet fever figures were markedly higher than in some surrounding districts.[123]

Haviland was to confirm this essentially optimistic view of the town in his subsequent investigation into the events of 1875.[124] Uppingham came out statistically well on every count, as well as in comparison with its neighbour, Oakham, which lay outside his territory. Even when he compared Uppingham's figures to those of a group of the healthiest districts in the country, it was one of the safest – except where scarlet fever, measles and child mortality were concerned. He concluded: 'These figures prove incontestably that the town ranks high in the scale of health'.[125] This pattern of overall stability – apart from a small dip in 1875 – is confirmed by the Registrar General's Annual Reports. In a typical year only a dozen or so deaths took place in the workhouse; there were a similar number from diarrhoea and between 3 and 7 fatalities from typhoid each year from 1870–4. The proportions of young and old who died seem unremarkable for the time. Over the period 1870–4 the burial registers show that 1870 was a hard year – 45 of the 61 deaths were of people aged under 18.[126] Thereafter totals were lower, and the proportion of young victims dropped to between a third and a half. The quarterly returns for spring 1875 show that 19 of the 36 deaths were of people aged 60 or over, and in the year as a whole, of 55 deaths 17 were under the age of 10 and 21 over 60.[127] Births in these years were remarkably constant, at around 145 per year. Moreover, Uppingham had not been singled out for inclusion in the special appendix of inspectors' reports nationwide from the years 1870–3, which covered 149 typhoid outbreaks, nor in the 20 special reports on such epidemics brought together in one publication in 1877. There had been no reason to include the town in the list of 42 inspections carried out in 1873 in connection with the local administration of sanitary laws.[128]

Haviland did, however, find one cause for concern in his first (1874) report. Uppingham's death rate from fever over a period was not falling as fast as in other places, after the good years of the previous two decades. During the 1850s only four of the twelve union districts in his area – Oundle, Uppingham, Market Harborough and Towcester – had had fewer than 60 deaths per 100,000 people through various fevers. The average for the whole of England was 88. In the 1860s (average again 88), only three achieved this – again including Uppingham, but its average figure had gone up from 45 to 54. In 1873 it stood at 57. He concluded: 'This is too stationary to be satisfactory'.[129]

Thus in 1875 both the Registrar General's Annual Reports and the findings of Haviland himself suggested that the town of Uppingham was better than many communities of its type in terms both of general disease incidence and of fever itself. It was also no worse, and possibly better, than many comparable towns in respect of sanitary and water-supply arrangements. However, as its population grew, the challenges facing its guardians were daunting. With both sanitation and water supply failing to keep up with this growth, there was plenty of scope for epidemic disease to take hold.

Haviland's work has the air of a man of zeal and energy, motivated by a passionate belief in the cause of public health. It was unlikely that he would have found a kindred spirit in the unambitious and somewhat defensive Dr Bell. The town was well provided for medically, and it is likely that its GPs, especially Dr Bell, were at the fierce end of the medical market spectrum. Bell was of comparatively modest medical calibre; by contrast, Haviland was very much the exception to Hamlin's view that many of the new MOHs had merely been promoted from lesser positions in the public health inspectorate. Uppingham represented only a small part of his territorial responsibilities: he was an energetic, busy, highly determined and quick-thinking MOH with a very different set of priorities from Dr Bell. This posed yet another potential problem for the RSA if it had to work with the two of them, on top of its prickly relationship with Thring and the school.

Neither doctor had, nor could have been expected to have, a completely clear idea about the causes of typhoid in 1875. This shaped their actions, but in contrasting directions: Haviland drove on hard for improvement, whereas in Bell's case his laissez-faire approach would come to haunt him.[130] It would also – along with the inability to carry out effective water analysis – make any definitive assessment of the causes of an epidemic hard to achieve.

4 TYPHOID: THE FIRST TWO OUTBREAKS 1875

Epidemic disease was by no means a new experience for Uppingham. The town had suffered plague in 1840, 1848 and 1850,[1] and there appears to have been a small outbreak two years later, followed by a major and more documented one in 1853–4.[2] The *Stamford Mercury* reported in December 1854 that 'the filth from the backyards which in most cases flows in to the open channels in the public street will scarcely be tolerated in any other decent market town'.[3] However, the paper had also carried reports four months earlier of

> a plan ... for a main drain at a depth of 10 feet that met with a good deal of opposition on the grounds that such a depth would drain the wells as well as the cellars, and the plans were accordingly rejected.[4]

In many ways this early skirmish between the RSA and local community serves as a symbolic event for the two decades before the 1875 epidemic, which was a period in which the RSA struggled to balance a local desire for improvement with fears about both its cost and the practical implications for householders. The LGB and its predecessors had numerous dealings with the RSA in this period – confirming the perception of people in Oakham in the 1860s that Uppingham's guardians were more proactive in sanitary reform than their own.[5]

In Uppingham the concerns – and the desire for improvement – soon re-emerged. In 1857 churchwarden William Compton, who had a house in the High Street and a wine and spirit shop in the marketplace,[6] complained about the state of the drainage. As a result there was an enquiry, and the Nuisance Removal Committee commissioned a survey of drainage options,[7] which resulted a year later in the main sewer being laid along the northern part of the town at cost of £750. This main pipe and its branch sewers covered much of High Street West and School Lane (including several school boarding houses), High Street East and Orange Street, North Street, Queen Street and Adderley Street. It ran down Seaton Lane to a sewage farm a mile from the town. However, it was laid at a very shallow depth and the diameter of its pipes was narrow (mostly only 9 inches).[8] Not all the properties in the streets it served were linked up to it. Seven years later (1865) the decision was taken to pave the streets with York slabs, at a cost of £1,101.[9]

Improvements raised expectations, but also created anxieties. The growth of the school had greatly increased pressure on the town's essential services, and the housemasters became increasingly worried about the lack of a proper water supply. One of them, the widely-read Theophilus Rowe, gave a lecture in the town in 1870 on the topic of 'Our Water'.[10] 'He showed that the water over a great part of the town was bad', recorded Thring. 'Mr Foster, a gentleman in the town, Mr Mullins [housemaster of West Deyne] and I had an Enquirer down, but the affair came to nothing.'[11] However, a year later (1871), with the debt for the earlier improvements paid off, an enquiry was held, at which the school made demands to the RSA for piped water and a better public sewerage system.[12]

In response to this call an LGB Inspector, Mr Pidcock, produced a far-seeing and ambitious report. Whether he came at the invitation of the guardians or on his own initiative is not known, but he pointed to the essential contradiction between trying to provide a water supply based on wells, which was being put at risk by the drainage measures already put in place, and an efficient system of sanitation based on water closets and cesspit drainage, which itself necessitated a good water supply, and which would be prone to contamination from joint leaks in the pipes connecting water closets to pits.[13] Pidcock recommended the abolition of cesspits and water closets in favour of earth ones. Proper rainfall channels and drains to convey drainage water away from the wells were highly desirable. He made a particular point about the lack of ventilation of the existing drainage, and became convinced that the small sewage farm to the east should be extended, and (ideally) a reservoir provided to the north of the town, to meet its growing size and needs. He estimated the total cost at about £6,000 plus any necessary land purchase.[14]

The school welcomed this report.[15] Wales, who was chairman of the Sewer Authority under pre-1872 arrangements, called a meeting of ratepayers to consider it. The meeting adopted a less ambitious scheme, however: the sewers would be extended, but it was decided to delay the installation of an intercepting tank and other extensions until the route of a proposed new railway was known. An application was made to the LGB for an £800 loan for sewerage improvements, to be paid for by a rate increase.[16] Messrs Whitaker and Perrott, an engineering company, was commissioned to produce a ten-page specification for extensions to go along the south side of the town, linking up with sewer pipes from the rectory and the market square. The new main sewer would be deeper and larger in diameter than earlier works (12 inches minimum, and 15 or 18 inches maximum, beneath areas of high density housing), and there would be frequent ventilators. This new sewer would run from the west of the town along Stockerston Road and past the Lower School, before heading south-east along South Back Way and across the London Road, alongside Ingram's field and thence to an extended sewage farm on Seaton Lane. Thanks to favourable gradients, very little pumping would be needed.[17] Another LGB inspector (Mr Morgan) reported favourably on the scheme in November 1871, including the additional cost estimates – £820 for piping, and £320 for disposal costs: £1,140 in all – and loan arrangements.

This episode shows very clearly why the impending Public Health Acts were needed to streamline sanitary law. A dispute grew up between the Sewer Authority (headed by Wales) and the Nuisance Removal committee of the RSA (headed by Barnard Smith) over precisely how the extra sewage would be deodorized. Wales's group was responsible for the proposed improvements, but Barnard Smith believed it would be unwise, possibly illegal, for the new sewers to be built before the water supply problems had been sorted out. He was all too aware that the main sewer still discharged into a ditch at one point along Seaton Road – a situation which local people had long complained about and which might actually cause pollution of wells if the sewers were extended.[18] When rainfall was heavy, the rising water table would put the wells at risk. Whitaker and Perrott drew up additional plans in March 1872. Despite the fact that there was smallpox in the town in June followed by scarlet fever in November, action followed only slowly.[19] There was a prolonged correspondence with the London authorities, after which there was another £400 loan application later in the year towards further sewage outfall works on the north-east side of Seaton Lane, which would now cost an additional £500.[20] The estimates were to prove optimistic; the tenders came in markedly higher than anticipated.[21]

Under the 1872 Act, sewer powers were passed to the new RSA, and Wales and Barnard Smith effectively joined forces. Both served on the new sanitary sub-committee, along with eight ratepayers who included two of the masters owning the largest houses (Hodgkinson and Rowe).[22] This sub-committee soon experienced opposition from the ratepayers, both to the rate levels needed to repay the loan and to the prospect of increased domestic costs to abolish their own cesspits.[23] At this point the issues widened. Sir Charles Adderley began to dispute both the siting and the lease arrangements for the proposal to extend the sewage farm, on the grounds of nuisance. His land lay next to the proposed site, which was owned by Wales. Adderley became involved in a dispute with one of his neighbours, John Pateman, a solicitor partner of William Sheild but, unlike him, a strong supporter of the school,[24] who favoured the proposed site. The argument centred on the extent to which the growth of the school was responsible for the increased pressure on the town services.[25]

Morgan came back to investigate further for the LGB, concluding that the school's 'villa residences' were the main cause of the changing nature of the western end of the town, and that yet further additions to the sewage farm works costing £120 would meet most of Adderley's concerns. It was agreed that the tanks should be covered and the ground planted;[26] Adderley then withdrew his objection. The work was carried out at the sewage farm, but taken as a whole it represented only partial improvement in so far as individual house drainage was left largely untouched.[27] Moreover, the seeds had been sown for future disputes between town and school.

As the costs of these works rose, other loan applications followed: £400 in 1873; £400 again in 1874.[28] The total sum was secured with funds borrowed from the PWLB, at 3.5 per cent interest. However, this was spent only in a piecemeal fashion on main sewers and a few extensions (for example along South Lane and Stockerston Road), as well as some outfall work. The LGB was concerned at the increasing cost,[29] and the PWLB was slow to process the loan. In January 1873 W. H. Brown, the RSA clerk, tried to persuade the PWLB to agree a loan over 50 years instead of 30, although he asked that the interest rate would still be at 3.5 per cent. The Board was not sympathetic: Treasury-driven retrenchment was starting to bite.[30] Further delays followed; Brown wrote again in March,[31] and was told that the LGB could not tell the PWLB how to conduct its business.

Meanwhile the guardians had repeatedly asked for guidance about good practice, and about the extent of their powers and status under the new Act. The LGB answered a whole series of queries between August and December 1872 about what could be delegated to sub-committees, the keeping of accounts, the loans already agreed, the discharge of debts incurred by the former sanitary authority, the terms of their clerk's appointment, the type of person they were expected to appoint as the new inspector of nuisances and MOH, and even about such detailed issues as the disinfection of workhouse clothing and bedding. Wales and the RSA then presented the Board in August 1872 with the first of a number of requests for the guardians to be given the status of a USA.[32] As a mere RSA, under the new Act they had powers over water supply, sewerage and drainage, nuisances and hospitals and cemeteries, but not over town improvements (e.g. streets and markets), lighting and the regulation of traffic. They would also complain repeatedly in the events to come that they had insufficient power to maintain and cleanse streets. USA status would also have given them the power to levy a 'general district rate' on agricultural land, to pay for increased responsibilities.[33] There was further correspondence on this issue in October 1873, at which stage the LGB believed that Uppingham was too small for such status – although it asked for a plan of the union boundaries.[34]

Barnard Smith, as chairman of the new RSA, struggled to keep control of its administrative costs. In May 1873 the District Auditor, Robert White, recommended that the union needed an additional collector of rates. In a note to the LGB he stated:

> Seeing the great improvements which have been effected in this Union owing as I believe mainly to the constant supervisions, untiring energy and good judgement of the chairman, I should think it unwise as well as ungracious to press the proposed

appointment against his wishes – although I confess my own opinion of the desirability of such an appointment remains unaltered.[35]

He may well have been correct in his view that the RSA's workload was running ahead of its resources; the clerk, William Brown, complained soon afterwards that pressure of work had caused him to submit land returns late; the LGB retorted with concerns about inadequate accounts records.[36]

Little respite came for the LGB in 1874. Its advice was sought over further audit problems, and over procedures for the appointment of a new master of the workhouse. Brown had to approach the PWLB again for yet another £400 repayable over 30 years,[37] this time for a dry earth storing shed which the inspector stated was 'virtually part of the sewage works'. Again an estimate had proved over-optimistic, and Brown had to assure the LGB that ratepayers in the other parishes would not be affected. Only two or three ratepayers turned up to object when the inspector came down to see things for himself, but it led to further requests from London for a breakdown of expenses.[38] Brown was conscientious but very persistent in his demands. There was a further stream of queries to the LGB about the salaries of a number of union officials and various audit matters, and about how far paupers' children could be made to travel to school.[39] The LGB also agreed a £2 payment to a local doctor following his attendance at a difficult birth at the workhouse.[40] By late May the guardians were again pressing for urban authority status – quoting a variety of earlier legislation as well as precedents elsewhere.[41]

Further research by Brown led to a long petition in July,[42] citing the need for stronger powers over offensive trades, slaughterhouses, rubbish clearance, the use of the marketplace, the keeping of animals and (significantly in the light of later events) the cleansing of footways and pavements. The guardians requested powers over the whole parish and not just the town itself; the LGB replied agreeing to an enquiry on 6 October. Posters advertising it were to be attached to the doors of all Uppingham's churches and chapels. Once again only two ratepayers other than the RSA members themselves turned up. The Board recommended that USA status should *not* as a whole be granted,[43] although certain equivalent powers to that status would be given in specific areas. Brown acknowledged this reply on 12 November, and proceeded to draw up a draft of revised bye-laws on the issues already listed, and on minimum space and construction standards for future new housing, the drainage of new streets and any upgrading of the present sanitary facilities.[44] These included detailed requirements on ventilation, foundation footings, damp courses and waste water drainage. The proposed bye-laws were submitted to the LGB early in 1875, but no reply had been received by 21 October when typhoid broke out in the school and Haviland wrote to the LGB:

I am at present much engaged in the investigation of a serious outbreak of typhoid fever in the school at Uppingham, and I find that the byelaws of the Urban Authority have not yet been returned approved. I take the liberty of asking you to expedite this matter as it is highly desirable that the said Authority should be able to exercise all the power it possesses.[45]

As 1875 began, low-level disease and public concern about sanitation were continuing features of life in Uppingham. Both the local inhabitants and their guardians were aware of the risks of water-borne disease, even if they could not explain the precise scientific processes that created those dangers. Debate and concern in this period centred largely on germ-theory causes, rather than threats from miasma or contagion. This was a year of weather extremes in England, with unusually large variations of temperature and a great deal of rain. Sharp frosts at the beginning of the year caused new cracks in sewer drains and cesspits, and deepened existing ones. However, this damage would not show up immediately, as the spring quarter of the year was one of the warmest and driest for nearly half a century.[46] Dramatic rainfalls then followed in early June – over eight times the normal level for that month[47] – making the town a sea of mud. Temperatures plunged again on 11 June, ushering in an early summer cold snap. This lasted right through to August, when six weeks of very warm weather set in. Autumn was a time notorious for typhoid outbreaks all over the country – especially when the weather was both mild and wet[48] – and in late September 1875 the rains and mud returned with a vengeance. This torrentially wet period caused a sharp jump in deaths amongst the elderly right across the land.[49] There was then a period of bitterly cold winds from 20 November to 16 December. By the time the mild weather returned just before Christmas, the school had long broken up.[50]

The pattern of weather extremes would continue through the first three months of 1876.[51] The two wet periods of the year caused annual rainfall at Rockingham Castle, a few miles to the south, of 37.4 inches. This was nearly twice that of the two previous years, and the highest for at least a decade – figures confirmed by Sir John Fludyer who kept records at Ayston Hall a mile to the north.[52] There was thus a classic pattern of wet weather coupled with typhoid appearance and reappearance in Uppingham during these months.

Early in February 1875 Thring recorded in his diary that there was 'much illness in the town – scarlet fever. This is anxious work. I fear we shall not escape. Most certainly in former days this perpetual fear of epidemics was not on schools as it is now ....' A few days later (8 February) he expressed further worries: 'We have sent in a memorial to the guardians requesting them to have the water analysed with a view to getting a proper supply for the town'.[53] He no doubt recalled the diphtheria outbreak in 1861, and it was only a few months since 'the sanitary inspector had told [him] that the town was in an uproar because he had taken a private house in which to treat cases of measles'.[54] He met the local inspector on the following day and the masters sent a petition to the RSA,[55] drawing its attention to the scarlet fever and the recurring presence of measles. The school also expressed concern about well-water pollution, both from house cesspits and from local animals: 'after every heavy rain the contamination from the various accumulations of filth on the soil is apparent in many wells ...'. It pointed out that Professor Attfield of the Pharmacological Society of London had analysed several samples (presumably on the school's initiative), finding that the water was pure on entry into the town but became speedily contaminated thereafter. A mains supply was essential.

By 13 February there had been four deaths from scarlet fever in the town in ten days. Thring wrote of 'anxious work', but also that:

> God has given me back some of the old elastic work power. In spite of this ... worry etc, I feel so full of life and spirits that I hardly know myself. I can do ten times as much as I have been able to do for years, and I feel cheery in proportion.[56]

A fortnight later, however, he was dejected again:

> Received an anonymous letter yesterday denouncing the filthy state of the town, and in a half-sneering, half-real way telling me to look to it, as no one else would. But I don't see how it can be done. The law helps us very little, and like most weak laws is a better instrument of oppression than of help.[57]

The scarlet fever outbreak had also attracted Haviland's attention. We do not know whether he visited the town at that point; it seems more likely that he received a report from the local inspector, Frederick James. Haviland decided that the town infants' school was probably the source, and he recommended closing it temporarily for thorough disinfection.[58] Thring kept up the pressure for further investigation, but little had been done elsewhere in the town to prevent a recurrence before the summer term began on 5 April – although the RSA agreed to send twelve well-water samples from various parts of the town to Dr Thudicum of the Medical Department of the Privy Council.[59] Thudicum reported back in July that all except one of the samples were heavily contaminated with animal or human sewage, and that the water was 'excessively hard and very unsuitable for domestic purposes. If Uppingham could obtain conducted soft water, it would do well to close up wells 2–12.'[60] It seems that the RSA, which made no response to this, was already afraid of ratepayer anger if it moved too fast.[61]

Meanwhile, on 7 June, a pupil in the Lower School (Hawke junior, aged 9) wrote home that he had a sore throat. His father was ill at the time and his mother could not leave, but she wrote to Mrs Hodgkinson, the housemaster's wife, asking her to see the boy at once. Although Mrs Hodgkinson hoped that it would be only a cold, she realized that it might be the beginning of a fever. She replied reassuringly that master Hawke was improving and now playing with

other boys again, but her husband wrote again on 17 and 19 June with news of
stomach problems and gastric symptoms. Lady Hawke came to visit her son on
21 June, and quickly realized how grave his condition was. She summoned Dr
Paley, a specialist from Peterborough. Young Hawke rallied, but then suddenly
collapsed and died on the evening of 24 June, as the shcool broke up for the summer. His death was certified on 28 June by Dr Bell as caused by enteric (typhoid)
fever. Bell was later accused of having failed to recognize the true cause of death
until he consulted a colleague, a slur which he rejected as being 'utterly at variance with the truth'.[62]

Although running legally separate institutions,[63] Thring and Hodgkinson
collaborated closely. It would be surprising if Dr Bell was not also consulted, but
they did not formally notify the RSA. They were under no obligation to do so,[64]
and they probably underestimated the danger which typhoid posed, hoping that
it was an isolated case and that, with nearly every pupil and many of the school
staff leaving the town for the summer holidays, the source of the fever would
vanish. With luck, it would all disappear over the holidays. However, Hodgkinson himself was ill for several weeks over the summer,[65] and it is possible that he
too had typhoid symptoms. He would later admit:

> I have to plead to ignorance of the nature and origin of typhoid fever. During my life
> of twenty years ... I had never had a case of this fever in my house, nor even seen one.
> There was nothing in the early stages of the disease to awaken anxiety in the mind of
> one inexperienced in the subtle forms it assumes.[66]

All this would explain later criticisms levelled at the school that nothing had
been done to investigate the origin of the outbreak for nearly four months after
Hawke's death, and until dramatic circumstances forced it to do so in October.[67]
Possibly the school also feared that unfavourable coverage in the press might
lead to a drop in numbers.

It is impossible to know the source of this first outbreak, although the exceptional weather was probably a contributory factor. However, while concern had
hitherto centred on risks from water and mud, the early autumn brought an inci-
dent which might well have seemed to support the views of miasma theorists.
On 2 September just before the new term began, a local plumber, Mr Chapman,
who lived in the High Street just a short distance away, was summoned by Hodgkinson to the Lower School. According to Haviland's later report – a version of
the event hotly disputed by both Hodgkinson and Thring himself – Chapman:

> ... was called in to remedy an obstruction in the flow of sewage from the boys'
> trough closets into an unventilated cesspit. The corner of the chamber in which the
> obstruction was supposed to exist being dark, a lighted candle was used, and almost
> immediately after it had been lowered on a level with the junction where an opening
> had been made, a tremendous explosion took place, the sewer gases igniting, passing

up to the ceiling like a streak of lightning, and at the same time burning the whiskers, eyebrows and hair of Mr Chapman.[68]

Coincidentally, this event came a week after the *Lancet* had carried a report of typhoid amongst 'men exposed to sewer gas'.[69] Only three weeks later, with term now well under way, a thirteen-year-old boy, Kettlewell, was taken ill with fever on 21 September, again in the Lower School. He was confirmed a day later by Dr Bell to be suffering from typhoid. Another case, again in the Lower School (Hastings major), followed on 28 September, and there were two more on 1 October. We do not know exactly when Thring was told, but on a day which he came to speak of as 'that fatal fourth of October', he recorded in his diary: 'Two or three cases of low fever in the school. This begins to make me anxious.'[70]

On Thursday, 7 October, another boy, Richardson, developed symptoms. His was a serious case from the start, and one which was to prove fatal.[71] Dr Bell was now very fully involved; over the next five days he saw up to ten other Lower School boys – along with eight boys from different school houses, and eleven other adults and children – mostly the families or house servants of members of staff). Some had very indeterminate symptoms, but Bell was fairly certain that at least two of those whom he had seen were also typhoid cases.[72] Only now – presumably after advice from Bell – were boys from the Lower School sent to the sanatorium rather than being cared for within their house. They had no automatic right of access to the senior school's sanatorium facilities. It might have been argued too, on the basis of any typhoid theory of that time, that it was better to confine the victims to their own living quarters rather than sending them to a building which was visited daily by older pupils from all the school houses and where they might pick up, or spread, this and other infections – either via contagion or through infected water or air there.

The school was now facing a deepening crisis – which seemed to be symbolized by the weather on Saturday, 9 October, when a football match took place between the pupils and an invitation fifteen selected by one of the masters (Uppingham football, like modern rugby, was played by teams of fifteen players). There had been torrential rain for much of the previous three days;[73] it was an exceptionally wet autumn. At lunchtime the rain began again. The *USM*'s correspondent braved the downpour to go in search of the captain of football, whose boarding house was some distance to the south of the main part of the school on the Rockingham Road. When asked whether the match would go ahead, the latter replied: 'We play through thunder and lightning'.[74] Having returned to his own house to get changed, the writer set off into 'the pitiless rain'. Eventually the game began – and a sizeable crowd of spectators braved the elements to watch. Afterwards:

> ... hot fires, hot water, hot coffee and alcohol in more than one shape were brought to bear on the soaking effects of the rain; and so well did they do their work that only

one member of the Fifteen, we believe, suffered from the game; and he got nothing but a cold.

Far from abating, the rain got heavier and continued for the rest of the day and the greater part of the night. The town was awash with mud and there were miasma concerns, too:

> The well-known malaria called the church-yard smell, which is almost as offensive as disinfecting powder, and must be a perpetual reproach to all ante-cremationists [*sic*], had thoroughly pervaded the atmosphere of the valley.[75]

After the game, the teams dispersed to various boarding houses all over the town. We do not know whether or not Thring turned out on the touchline. He had been in post for over two decades, so he would have known all the members of the visiting team; he would surely have been there in normal times, but he had plenty to worry about, with illness in several boarding houses and a number of private properties. On the Wednesday before the match, the weather had improved and he had hoped a dry spell would chase the illness away. Thring had recorded his fears in his diary on the Thursday about the condition of the ailing children of two members of staff: 'All this presses very heavily and makes one nervous about one's own children too'. A day later he had written: 'The bell tolled this morning [in the town], and I was in great fear, but a man had died in the union [workhouse]. I very much fear that we shall not escape death.'[76] Six boys had been admitted to the school sanatorium on the day of the match – joining seven others who had been there for between one and five days.

The evening of the day of the match also saw the arrival of a young man who was to become a chance casualty of the typhoid outbreak.[77] A coach drew up outside the Falcon Hotel in the marketplace; it had met the train at Manton station and had then travelled the few miles over to Uppingham. A seventeen-year-old passenger stepped down from it. We do not know his name, but he had come all the way from Southampton, having agreed to take up work as a pageboy in the Lower School. The school was later to claim that he was offered his fare back to Southampton but chose to stay.[78] Haviland would dispute this, alleging that the boy replied: 'If I had known, I would not have come; and if I had money in my pocket, I would go back again'. Either way, just over three weeks later he would be dead.[79]

On Sunday, 10 October, the day after the match, the chapel service raised Thring's spirits,[80] but there was a steady stream of new cases in the days which followed: a few milder ones in the town, but mostly in the school. There were five in West Deyne, two doors to the east of the Lower School – all involving boys aged thirteen to sixteen,[81] as well as young Cecil Mullins, the housemaster's son. The baby son of Paul David (the school director of music, who lived in a neighbouring house)[82] was also gravely ill. At the Lower School, Hastings's younger brother had gone down with the disease. More worrying still was the case of Stephen Nash, aged fourteen, of Redgate – a house on London Road nearly half a mile from the other affected ones. There could still have been any number of ways in which boys travelling around the town might have ingested foul water, especially at a time of intense rain. However, it must now have seemed more possible that contagion should be considered, as boys from the various houses came into contact with one another in school – and perhaps spread the impact of something brought in to town and school by an outside source or carrier over which the school had no control. This geographical spread of cases made it far harder for all parties to determine with any certainty whether they should be looking for germ (i.e. water-borne), miasma and/or contagion causes of the epidemic

Nash had complained of feeling faint during singing practice; Dr Bell saw him that evening and again two days later.[83] Thring met Nash's parents when they came to visit their son in the sanatorium, and described them as 'kind and sensible'. But alarm began to spread amongst pupils and staff, and rumours started to reach parents. A number of them reacted with aggression or panic, demanding action and then starting to withdraw their children. Some arrived unexpectedly at the school: at least two mothers and one father were at the bedsides of more serious cases.[84] It is likely that even the mildest cold symptoms amongst boys would cause alarm to them and their friends – and that some imagined they were ill, as this anxiety grew. Local rumour soon put the number of cases at nearly forty, although Dr Bell – to whom Thring remained staunchly loyal – insisted publicly at this stage that it was probably only just over a dozen.[85]

On Monday, 11 October Wensley Jacob – Liverpool businessman, father of two pupils and a school trustee – contacted Thring. Six other parents in that city, two of whom were doctors, had formed a deputation to see Jacob; they were demanding that the school summon the MOH.[86] The next day a letter arrived from Dr Grimsdale, another Liverpool parent, 'speaking in the name of many parents in a kind spirit, but also in an imperious one'. The rumour of forty cases was racing around Merseyside. Thring now faced a very difficult decision. If he closed the school and sent his pupils to their homes all over the country, he risked spreading the infection. He might accelerate the panic – with dire consequences for his school's reputation and future. It might never reopen. But if he kept it in session and the epidemic grew, he risked being accused of complacency and secrecy, and of putting the school's and his own interests ahead of those of his pupils. In the long run this might prove even more damaging. As headmaster it was his main responsibility to prevent a sense of ever-deepening crisis, amongst staff, pupils and parents. He felt on balance that he needed to keep school life going on as normally as possible, even if some of his housemasters and other staff showed signs of panic, or faced personal tragedy within their own families. He also needed to summon up the right mix of assertiveness and tact in dealing with an RSA which he believed increasingly to be complacent and supine.

On Tuesday, 12 October, Thring, backed by Mullins, the West Deyne house-master who was facing increasing cases amongst his own house and watching his own son deteriorate, decided that he had no alternative but to ask for help from Haviland. [87] There is no record of Bell's attitude to this. Thring wrote ask-ing Haviland to come over urgently from Northampton to 'test and examine' the drainage system and water supply of all the houses. Either out of a sense of courtesy, or possibly because he saw it as tactically sensible to adopt such a tone, he added:

> If you cannot come yourself, perhaps you would kindly telegraph to me, as it is no use to us to have the inspection by any man whose name will not carry respect and conviction, not only in this immediate neighbourhood, but also in other parts of England amongst the parents of the boys, in the great towns especially ...[88]

But the pressures continued to mount:

> I had scarcely finished breakfast when two mothers came in to get an arrangement made at [the sanatorium]; then the masters' meeting; then in school an interview with Bell about [sanatorium] arrangements.

He had meetings with housemasters and other staff in and around the school:

> ... met Christian [housemaster of Redgate], coming back, who said Nash was [thought] to be dying, Mullins had already told me that there was no hope for his [own] little boy; wrote part of another letter, went to dinner, lay down, but was sent for by poor Mullins. I found him quite perplexed about his house, overdone both in body and mind ...

In an attempt to work off the pressure, Thring went for an hour's walk that day with his youngest daughter, Grace, and then spent time praying. Briefly there was hope that Nash and Cecil Mullins might be rallying. But by Wednesday things were bad again – and now Hodgkinson needed support: 'driven out of his wits by the calamity and fuss. I very much fear that he will not stand it ....'[89]

Thring decided that he would not send the pupils home, and he spent part of Thursday morning (14 October) addressing his staff:

> ... under no conceivable circumstances should I break up the school; that it was a great injustice and wrong to many forcing them to have their boys home; that, in the first instance, when a house was at all got hold of by illness, I should have parents written to, to be told the fact, but strongly dissuade the removal of the boys; then if it spread I should make removal optional, and if it got very bad should throw the responsibility of keeping them here on the parents. That we should always stay so long as there were any boys to teach and keep them.

He emphasized the dispersed nature of the school's layout, with houses all over the town:

> I said we were not in a big barrack with one common establishment, but each with his own house and separate arrangements, which renders it quite unnecessary to break up the school. I also told them I should not permit the school to be overhauled by any but a competent and true authority.[90]

We do not know when the guardians individually became aware of the full extent of the crisis, but in a community as small as Uppingham they must by now have known that the situation was bad and growing worse. Before their weekly Wednesday meeting Brown, as their clerk, had instructed Mr James to investigate. At that meeting James confirmed formally that there were typhoid cases in the school. Keen to be seen to be as proactive as the school, the RSA members decided, like Thring, that this was a matter for Dr Haviland. James was instructed to send a telegram to him: 'Fever in the school houses here; your immediate attendance is requested; a committee will be summoned to meet you; reply by telegram'.[91]

Haviland was elsewhere at the time, but he returned home on Thursday to find these communications from both school and town. He immediately replied that he would come over the next morning. There was also a strained meeting between Mr James and Thring, at which accusations of secrecy and inertia were traded. Thring wrote later:

> Was not a little amused to hear from him that he [claimed to have] known nothing of any fever in the town until today at the Board. So I may be excused for having known nothing [about illness in the school] ...[92]

Haviland's impending arrival, and further demands for information from Liverpool, weighed heavily on Thring. He was particularly concerned about the Lower School: 'I really fear it will send poor Hodgkinson into his grave ...'.

On Friday morning Haviland arrived from Seaton station at 11 a.m. to begin enquiries.[93] At much the same time the disease claimed four year-old Cecil Mullins, who died at West Deyne. Another telegram came from Liverpool, demanding to know whether or not Haviland had started his investigations. 'When will it end?' wrote Thring in his diary. 'I am myself very tired and done up ... all one's feelings of joy in doing one's best, and the happy sense of unselfish working and pouring out of liberal free life on one's work is so utterly destroyed ...'.[94] The achievements of two and a half decades might now 'melt like the snow of spring'.[95] Thring attended Cecil Mullins's funeral on Saturday, 16 October. Sunday brought the death of Richardson. Nash from the Lower School followed on 21 October and Oldham, another Lower School pupil, who had been in the sanatorium for only 24 hours, two days later.

The list of those in, or connected with, the school who were affected by fever of varying degrees of severity did indeed come eventually to over forty. It included no fewer than seventeen from the Lower School, and nine from West Deyne.[96] Six different senior school houses were affected. The sanatorium list includes crosses against the names of four marked as 'an undoubted case of typhoid', although we cannot be sure when these crosses were included.[97] Dr Bell also recorded that twelve of his patients in the town (as opposed to the school) showed similar fever symptoms during September and October. They included two children of John Hawthorn, the bookseller, and H. H. Stephenson, the school's cricket professional, but none of these town cases were serious, and all would recover.

We do not know whether Thring and Haviland had ever met before. It is likely that Thring was more than keen to enlist his support; Haviland was well-placed to put pressure on the RSA, and he possessed the necessary experience and authority to call on the LGB to add its weight if necessary. This explains why Thring's telegram asking Haviland to come over from Northampton had laid emphasis on the latter's professional expertise. On Saturday, 16 October, Thring met Haviland within an hour of attending Cecil Mullins's burial in the church-yard barely 150 yards from Thring's own house. Bell does not seem to have been present. At almost exactly the same time a group of angry parents gathered at the Falcon Hotel nearby. Feelings were running high; as another father arrived late for the meeting, those already there asked him whether he had come 'to take his boy out of the hands of these murderers'. When Thring heard about it later, he commented ruefully: 'Nice for poor old Hodgkinson, whose whole life has been bound up in the house and boys; nice for me too, for I am murderer No 1'.[98]

Haviland had wasted no time in looking around the town as well as the school. He was initially furious with the RSA when he discovered that the problems with the privies at the infants' school had still not been rectified, eight months after he had pointed them out.[99] At this stage, although already convinced that the typhoid had originated in the Lower School,[100] he advised that it was quite safe for the school as a whole to continue. Thring, who like many of his colleagues was probably coming across public health officialdom for the first time, found Haviland's manner hard to take, and was also worried that he might be listening too much to alarmist rumour about the spiralling number of cases:

> I confess that my blood rather boiled when I heard this man deliver an ex cathedra statement, as if all he said was gospel on a question where there was so much to be considered ... Had he decided otherwise, I don't know what I could have done. It was strange, too, to hear [him] fussing about the lies that had been told and following them up.[101]

A few days later, on 21 October, Haviland carried out a thorough inspection of Thring's own boarding house. Thring was more reassured this time: 'I am glad to say there is little to be altered, and that not of much consequence. He also passed both my wells officially as perfectly pure.' This was to be confirmed a week later with the arrival of results of water samples which Thring had sent for analysis to Messrs Savory and Moore in London.

Yet in the days which followed, Haviland's advice was to change a number of times. He sent Thring a telegram on 22 October saying that in his opinion all boys in infected houses should be sent home,[102] although he subsequently wrote to one housemaster (Christian) that he saw 'no danger whatsoever in allowing pupils to remain: the sanitary arrangements of your house are such as to warrant me in believing that Nash did not contract fever under your roof from any sanitary arrangements in your house'.[103] The fact that Christian's house was one of those furthest from the main part of the school may explain this apparent contradiction. At this stage Haviland attempted no judgement about the source of the disease, nor between the various theories about what normally caused it. Unsurprisingly, Thring decided to act on the second message rather than the first; he was very reluctant to disperse the school. He drew up a statement for parents of the guiding principles which would apply in dealing with the crisis. Emphasizing again that the school consisted of eleven geographically distinct houses, each with their own catering arrangements, he reiterated all the reasons why he was unwilling to send the boys home – although if parents insisted on it, or if the disease became really established in any particular house, he saw no alternative.

He consulted his two trustee allies, Birley and Jacob, about the statement; their knowledge of the Liverpool parents would be important. They wondered whether other parents might think this statement too dictatorial, but Thring pressed ahead and later claimed that parents had praised it.[104] He also received a letter from Dr Christopher Childs, an old Uppinghamian who had recently qualified from St George's Hospital in London, offering his services to the school.[105] Childs had graduated from Oxford some years earlier with first class honours in science, and had been a very successful footballer and athlete whilst at the school. Thring decided to take Childs on to the staff as science master and sanitary officer. Recruiting new, additional staff at a time when many people were gloomy about the school's future would reassure parents and staff, and it would go down well with the Old Uppinghamians. This appointment might also relieve the pressures on Dr Bell – although Thring failed to anticipate what a source of friction it would prove to be when Bell later started to fear that his own position and income were threatened by Childs's presence.

We can only speculate about precisely what determined the RSA's action in this same period. Some of its members were at best lukewarm towards Thring and the school (for reasons already identified), and initially they would not have been overwhelmed with regret at his discomfiture. Even when the crisis deepened, and having joined the call for Haviland to investigate, it made sense in many ways for the RSA to await the outcome of his findings. It lacked the technical and medical expertise to ensure that any action in advance of Haviland's recommendations would be cost-effective (or, indeed, effective at all), and it is unlikely that they had any contingency funding for a crisis of this magnitude.

As the full extent and implications of the epidemic sank in, however, the RSA realized the importance of giving the impression of decisiveness and of being in control. Thring believed that its members were keen to place the blame firmly on the school,[106] and at an early stage it pronounced that wells in one boarding house on the edge of the town (Redgate) were quite pure, only to have chemical analysis of samples submitted by Thring to Savory and Moore sent back as 'turbid' and over-heavy in carbon and nitrogen.[107] On 27 October the RSA served notice on four of the masters to 'remove nuisances arising from their cesspits',[108] following initial visits by Haviland. Whether through germs, miasma or contagion, the RSA held that these pits were surely to blame in some way.

Thring's irritation at this action was compounded by the reaction of the school trustees, when they met on 29 October. They declined to get a sanitary expert down to give advice, preferring instead to pass a motion:

> ... recommending to the headmaster at once to close the school, and that a representation be made to the Sanitary Authority urging them to give every facility for the work pointed out as necessary to be done at the several school houses by the report of the Medical Officer of Health.[109]

They would set up a sub-committee to work with the RSA. Thring was particularly scathing about what he saw as this spineless response:

> A most bitter disappointment. The trustees with all this great school handed over to them ... have appointed a sub-committee to urge the sanitary authorities here (whom we mistrust and despise) ... it is very hard to keep down the bitter, sour feeling which this day has once again curdled within one ...[110]

There was no alternative now to announcing that term would end on 2 November while necessary improvements to the houses were made – even though, with the prospect of the school being closed for at least two months, parents would surely start to consider alternative schools. Thring was careful to promote unity amongst his staff at this critical time. A series of meetings established that, costly though it would be to the housemasters, the houses must be put into a sanitary state with which no-one could find fault. Parents were told that the school hoped to reopen 'the week after Christmas Day'.[111] Thring confessed in his diary:

The last evening, alas! of our maimed school-time. It is strange though, the childish relief I feel at not having to get up for school tomorrow. A true and real relief, however, is the lifting of that fearful weight of the possibility of fresh fever. For the first time for many days I have drawn something like free breath.[112]

The recriminations between school and town came fully into the open as soon as the boys had gone home. The RSA made public on 5 November its order to Thring and the other masters to 'remove nuisances arising from their cesspits'. It then went further: a resolution was unanimously carried that 'serious blame attaches to the masters in whose homes enteric fever originated'. It also criticized Dr Bell for the first time: 'The medical officer of the school was also blamed for not investigating the causes ... and reporting the outbreak to the Sanitary Authority ... for not attending a meeting of medical practitioners of the town, convened by the MOH ...'. Finally they decided to commission a notable sanitary engineer, Rogers Field of Westminster, to report on the drainage (both public and private) of the town.[113] Thring was deeply angry at this apparent rush to judgement before the evidence had been assembled and assessed:

> Wrote to Jacob and Birley, but could give them no information excepting that we were going to be made a scapegoat of. And sure enough they [the RSA] have been and done it. I got tonight such a document, the most wonderful bit of Jack-in-officism. Considering that the Sanitary [Authority] and town generally have steadily resisted all improvement as far as possible these twenty years ... and have ignored fever in the town for the last six months, they have had the audacity to attach serious blame to our houses for having had fever; no reasons given, no hearing of the case. It is astonishing. Altogether it is the most insulting thing I ever knew. It is truly laughable, but it is noxious too, as they mean to send it to every parent whose boy has been ill ... They think nothing can touch them. The inspector calmly told Guy that if we applied to the London Board, they would only send down the complaint to him, and he had better save himself the trouble. I shall have some difficulty in keeping the masters reasonably quiet under the insult ... Altogether this is a time of humiliation and sackcloth.[114]

With the RSA apparently intransigent, Thring no doubt feared that Haviland's full report (due within weeks) could well take the RSA's side against the school. An MOH's sympathies might well lie instinctively in that direction, especially if Barnard Smith could convince Haviland that the RSA had been improving the sewerage as fast as it reasonably could. Moreover, it was Haviland's job to protect the health and interests of the whole local population, not just the school. His priorities were likely to be very different from Thring's. It was at this point that Thring became convinced that he needed to enlist influential support beyond the immediate locality. He could get little sympathy or action from the RSA without it, and for the next few weeks his main aim was to force the LGB itself to become involved.

The list of obstacles to rapid sanitary reform in Uppingham in the pre-1875 period might have daunted even the most zealous and skilled Authority – although the suspicion remains that the RSA could have been more active over the scarlet fever outbreak at the start of 1875. Once the epidemic broke out, it rightly summoned Haviland as MOH, but having commissioned Rogers Field to investigate the outbreak, it then moved with questionably selfish speed (possibly reflecting a degree of panic), before he had done so, to heap all the blame on to the school, without any precise evidence as to the source. The absence of speculation by either MOH or GP about how the typhoid had been contracted might support the idea that knowledge at the centre did not always filter down quickly into the localities, that Haviland was less knowledgeable than he liked others to believe, or that, not yet having fully inspected the school and its houses, he was reserving all judgements until later. Water analysis provided little help as it centred on chemical, rather than microbiological, issues – confirming the fact that the *next* decade was the decisive one in this sphere.

Where the school is concerned, once typhoid broke out, it is perhaps understandable, with no obvious course of action to take over the summer, that Thring had hoped that time would solve the problem. When it did not, he faced a difficult choice over the issue of sending the pupils home – although his reluctance to do so becomes less surprising when the subsequent tentative stance of the trustees is taken into account. With the benefit of hindsight it is arguable that the school was at fault on two issues: it allowed itself to be too reactive, and thus open to charges of secrecy from both parents and RSA, and it permitted too much freedom of movement between houses – a charge which Haviland would have no hesitation in levelling at the hapless Dr Bell, even though it seems at odds with Haviland's apparent willingness to let Christian disperse his boys to homes all over the country. Events – and personal tensions – would now assume a momentum of their own.

5 WINTER 1875-6

On 5 November, Thring wrote to his brother, Sir Henry Thring, who had extensive influence in Parliament:[1]

> If you can get to Sclater Booth and the Central Board [i.e. the LGB], it is simply all in all to me ... if it rested with us, all could be set in order without difficulty. But it does not rest with us. The town is at fault ... Unless we can get the central authority turned on, it is pretty well ruin ... The town is trying to make the school its scapegoat, for the double purpose of hiding past mismanagement and preventing present outlay and exposure ... Uppingham may forget but cannot forgive, that it exists mainly by the school ... The Authority ... cannot act with vigour enough, and [even] if it could do so now, the row and panic amongst our parents is so great that it would not help us much after the lies and exaggerations that have been set going ... You government men have no conception of local tyranny.[2]

Thus the RSA gained a further cause for complaint against Thring: his willingness to put pressure on it by exploiting his powerful contacts to the full. By November he had become convinced that he needed to force the LGB to conduct a full enquiry. However, while such a tactic might strengthen the school's position, it also risked deepening the antipathy with the town. Indeed, the RSA came to believe that Thring was determined to divert all blame away from himself. Three days later he contacted Sir Henry again: 'I want nothing but fair play and no favour. Nor do my masters; they are honest and hard-working, and ready to do anything that is judged right.'[3] On receiving this second letter, Sir Henry went to visit the LGB at once. It now assured him that 'Government would do everything it could'.[4]

A second important contact was Sir Charles Adderley; he too was prepared to use his extensive political connections.[5] Adderley wrote to John Lambert of the LGB on 15 November. He pointed out that he was a large local landowner, and that while he recognized that there had been some improvements in the town in previous years, they were nothing like enough. He painted a bleak picture of a town whose sewers were inadequate and incomplete, of leaking cesspools and 'some filthy manure yards and pigstyes in the town, which must to a certain extent pollute the wells'.[6] He believed that the RSA needed greater powers to insist on their removal, but also that it had not made enough use of the byelaws which already existed.

Adderley believed furthermore that relations between the RSA and the school had deteriorated too far for much action to be initiated locally. He called for a government enquiry, lest good ratepayers' money was poured after bad: it would be no use adding piecemeal to an incomplete sewer system; the current ones needed to be re-laid more deeply. In his opinion these issues were more important than a new water supply, even though the wells were suspect. It is possible, however, having resisted the sewage farm development on his own land a few years earlier, that he was now keen to avoid the prospect of being pressurized to give other land for a waterworks.

Between them Sir Henry Thring and Adderley succeeded in the quest for an LGB enquiry. Headmaster Thring was able to write to *The Times* to that effect on 16 November. A trip to London to confer with Sir Henry Thring and Adderley did him a power of good:

> On Monday I came back, having been put in another world by this day's absence – lifted clear out of the old rut. And set with a stronger, clearer faith, on higher ground … Thus ended a great day for me. The local tyranny is now shut up for a time … Altogether I feel a great cloud rolled away, and begin to see light and breathe freely.[7]

It was essential to put the RSA on the defensive, and to bombard them with initiatives; the school must be seen to be the party most actively addressing the problems. To this end, Thring and the masters now engaged Alfred Tarbotton, a Nottingham engineer, to recommend improvements to drainage in the houses. In a letter to Christian on 4 December Tarbotton urged the housemasters, on whom most of the expense would fall, not to drag their feet over putting these measures in place.[8] Thring wrote to some of them on the same day, reinforcing this message.[9]

Thring was now determined to address not just the drainage question but the water supply as well. Two separate analyses (one in the spring and another in the autumn) had highlighted the contamination of the wells. From early November he was at work on plans to spend £100 on trial borings for a new water supply.[10] A private water company might well be the best means to carry it out, although such independent action would hardly endear the school to the RSA. Hodding and Beevor, a firm of solicitors based in Worksop, were consulted; they advised that an Act of Parliament should be sought during the next session to permit such a company.[11] The RSA responded predictably by giving notice of opposition 'merely to protect our own interests and those of the ratepayers'.[12] This notwithstanding, a week before Christmas the solicitors sent out the first draft company prospectus to Christian and the other masters, and trial borings started.[13]

Even though the battle for an enquiry by the LGB had been won, an analysis of its papers during November shows that this outcome could not have been taken for granted. The LGB's senior officials differed amongst themselves over the wisdom of becoming involved. This is not surprising, given the internal tensions within the Board itself and the remorseless lobbying in this period by both town and school. The RSA had been every bit as intent as the school on winning the LGB's support. It had also received from Haviland, very soon after his arrival in Uppingham in October, a warning that the epidemic looked serious, and support for the Authority's concern for a speedy decision over stronger bye-laws.[14]

The pressure exerted on the LGB by the school in particular was unremitting. Thring contacted it on 1 November. His allies amongst the school trustees, Birley and Jacob, were concerned that Haviland might not be able to give Uppingham his undivided attention, and they urged Thring to seek assurances that the LGB would intervene as necessary. One phrase in Thring's letter made the Board wary. The words: 'We recommend you to throw on the London authority the responsibility of any future outbreak' were highlighted by a civil servant as a phrase to be treated with caution.[15] Thring also made a personal approach to Sir John Simon – introducing Dr Childs, who would call on him to put the school's case for more urgent action, because it was worried about 'not satisfying public opinion'.[16] Childs duly appeared, but Simon, sensing that the LGB was in danger of becoming too involved in all this controversy, replied that he had great confidence in Haviland and 'it was not necessary to supersede [pre-empt] him'. He believed that the cause of Uppingham's problems was well known – although if 'the college' thought the town's drainage was inadequate, it could make a formal complaint under Section 299 of the recent Public Health Act, in which case the LGB would no doubt send down an engineer to assess things for himself.[17] He told Childs that such matters could also be referred to Rogers Field, an engineering expert who was now conducting an investigation into the state of the town for the RSA.

Simon's concern not to yield too quickly to the school's demands may indicate the influence of an encounter earlier in the afternoon of Childs's visit, when another Board official had had a conversation with Field. The latter, possibly out of a concern to defend the position of the RSA as his new client,[18] stated to this official that:

> He found the masters (or some of them) less anxious about the perfection of the sanitary arrangements in their houses than they were about doing by exact measure the [minimum of] work which would satisfy the sanitary authorities.[19]

Field also suggested that the housemasters were anxious about the costs of the improvements which Tarbotton might recommend within the boarding houses. The LGB memorandum about Childs's visit states that Field's comments 'were in private conversation, but I noted them in connection with Mr Thring's urgent letter brought by Dr Childs'.[20]

In making these approaches the school was effectively asking the LGB to bypass Haviland and to pre-empt his forthcoming report. It is hardly surprising that the Board declined to do this, preferring instead to play for time. It assured the school that there was no cause for undue alarm, and that Field and Haviland would probably recommend radical improvements to the town rather than piecemeal ones. It was keen to avoid taking sides too soon, and especially to avoid backing either party simply because it protested the most.[21] Thring greeted the news of Simon's distancing response with deep gloom: 'Childs cannot get us masters any help from the London Board for ourselves and we are quite at sea'[22]. Undaunted, however, he wrote again on 21 November. Childs would be visiting the Board a second time to put the school's case. This visit achieved no more than the first one; the LGB notes state that it reiterated the school's right to make a formal complaint. It added that Childs appears to have 'hinted, but was not able to speak positively' that the school doubted the sufficiency of the town sewerage.[23] Having only recently been appointed to work at his old school, he may have felt the need to express himself less forcefully than Thring, who had lived with fears of epidemic disease for many years.

Not convinced that these approaches were sufficient, Thring and the masters had also sent simultaneously to the LGB a formal complaint against the RSA. The complaint pointed out that the school paid large sums in rates; it needed urgent help and wished the LGB to send its engineering inspectors to Uppingham to see things for themselves. It was addressed to the president, Sclater Booth, himself – and therefore it appears initially to have by-passed Simon and those officials who had seen Childs, and who were aware of Field's criticisms of the school. As a result, the initial reaction within the LGB to this complaint was inconsistent with the line which Simon was already taking with Childs. The LGB's copy of the complaint is annotated (probably by Sclater Booth).[24] It took a much more urgent view of the school's plight than Simon had shown. The annotation includes the words: 'Will you deal urgently with this? It is an exceptional case, and I think we ought to appoint but one inspector and Dr Haviland ought to suffice.' 'JS' (presumably Simon) conceded that Robert Rawlinson (chief engineering inspector) might be sent, but urged that he (Simon) be kept in touch with events. He also warned that sending an engineer might be diplomatically difficult, when so many others (Haviland, Field and Tarbotton) were already involved in similar investigations. In the end, the LGB – keen, as ever, to be even-handed – sent information about all this to the RSA, with the request that it pass the information on to Haviland as well.[25]

During a week-long period from about 11 November, the LGB's officials debated whether it could stay out of this dispute any longer. There were those who believed it should remain detached: a Board official (Hugh Owen) wrote an internal memorandum to Rawlinson a day later that 'Mr Fleming saw [another] Uppingham deputation yesterday'. He thought Rawlinson 'might have been saved the trouble of visiting the place ... but the President has returned the papers this morning and he still wishes that you should make the inspection originally proposed as early as practicable'.[26] Rawlinson himself (not surprisingly, for one who had always adopted a policy of non-compulsion on local authorities) argued for minimal involvement. In an internal memorandum he wrote: 'I can only advise [that] after my visit ... it is important that you repudiate the idea of responsibility for any future outbreak. The responsibility is, and must remain, local.'[27] However, the combined efforts of the school, Sir Henry Thring and Sir Charles Adderley finally convinced the LGB on 15 or 16 November that it had to become involved. Rawlinson would indeed be sent down to Uppingham, to add his views to those already being collected. Sclater Booth, who had been less hesitant about central intervention than Simon and Rawlinson all along, took the final decision.[28] Simultaneously, there was a further, but inconclusive, exchange of notes between the LGB and the RSA on the bye-laws question.

At various points in November and December the LGB also received updated information from the school, the trustees, the RSA and various water analysts about their respective activities.[29] All its reluctance to become involved must have been confirmed by a communication on an additional issue. Dr Bell wrote to Sclater Booth on 12 November protesting about Haviland's actions. A private war was developing between the two doctors, which probably stemmed from Bell's own defensiveness about his past actions or omissions, and resentment at what he perceived as Haviland's overbearing conduct.

Bell had already complained to the Authority about Haviland's demand for information about his patients. He also resented Haviland's attempt to force him to come to a meeting which Haviland had called with all three town GPs – and the implied support of the RSA for such dictatorial action. The Board wrote back supporting Haviland's actions but stating that Haviland had no legal right to make Bell appear. Beyond that, it could not express an opinion on the RSA's actions in what was a local matter.[30] Bell persisted, writing again on 7 December in great detail. Far from being indifferent to Haviland's enquiries, he had met the inspector of nuisances Mr James, and he had talked at least three times with Haviland himself. He had had no time to attend the doctors' meeting, because it had been called at very short notice – which seems to be at variance with his and Haviland's other statements that Bell had pleaded sudden illness as the cause. If he was under fire for not reporting suspected typhoid cases to Haviland in June and October, he would like to know whether the other local doctors were under similar investigation over alleged cases in the town a year earlier. Haviland had allegedly visited Bell's patients unreasonably and repeatedly, and had also suggested alternative treatments for some of these patients.[31] Bell was to write at least three more times in similar vein. It is hard to believe that his personal struggle with Haviland helped the school's cause very much at this point. The LGB dutifully replied each time, and passed news of Bell's approaches on to Haviland for comment. Haviland replied on 23 December that his forthcoming report would clearly rebut such charges. The LGB did, however, note in its files that Haviland might have breached professional etiquette, even if he had not exceeded his authority.[32]

The battle had meanwhile entered the columns of the national press.[33] On 30 October the *Lancet* carried an editorial highly critical of the school,[34] probably fed by discontented parents whose sons had been taken home before it broke up on 5 November: 'In all the houses the cesspit system has been adopted, and the water-supply is in dangerous proximity. In some of the houses the dormitories are supplied with water from the cisterns which supply the water-closets.' Pointing out that, by contrast, the town was apparently free from typhoid, the writer suggested that there was much which Haviland should investigate within the school. The *Lancet* also carried a letter from an anonymous 'Medicus',[35] claiming to be a relative who had visited one of the stricken boys. It alleged that he had been given evasive answers to his questions by both Hodgkinson and Thring, and that the boy had turned out to be in a room at the sanatorium with one of his contemporaries 'in the second week of typhoid fever, with a temperature of over 105 degrees fahrenheit'. Dr Bell had allegedly tried to avoid meeting 'Medicus', who complained:

> ... of the school authorities for not telling parents when the fever first broke out, for not telling them when the children first showed signs of the disease, for studiously shutting their eyes to the real nature of the disease, for doing nothing to improve the sanitary arrangements and for keeping boys in places which were mere pest-houses, as I hear they have been condemned by the medical officer of health.[36]

Thring took the opportunity of a masters' meeting on the following day to insist that no-one should write to the press without consulting him first. He wanted to keep control of any press counter-claims himself. He may well have been anxious to avoid too much of a public slanging match, for he had just received a letter from Bell admitting problems in the sanatorium – of ventilation, a shortage of beds and lack of cooking facilities and storage space. It seems likely that Bell had also raised these issues with Mrs Grigg, its superintendent, causing her to resign.[37] Whilst having to replace her was probably an added inconvenience at the time, it may in retrospect have been seen as fortuitous. Forty years later in an obituary article on her successor (Miss Goodwin), the *USM* stated that the new matron had completely reorganized the sanatorium on her arrival[38] – a remark which also gives some credence to the criticisms which Haviland would shortly make of the sanatorium arrangements at the time when typhoid had struck the school.

The Times took up the attack on 5 November, quoting from the RSA's hurried statement of self-defence,[39] which had just been published. It lambasted the sanitary arrangements at Redgate, Christian's house, whose boys had included the late Stephen Nash:

> It appears that the house is quite isolated and has no connection with the general drainage of Uppingham – in fact, all the sewage is received into cesspits which are close to the water-supply. The cisterns supplying the water closets in nearly all the affected houses are furnished with taps from which the dormitories (too) are supplied with water. Sinks both above and below empty themselves directly into the drains; and in fact it would be impossible to find arrangements more directly fitted to engender and spread the special disease which has shown itself at Uppingham School. Though the house in which the fever first showed itself [i.e. the Lower School] is a splendid mansion, the architect seems to have altogether forgotten to provide for the health of its inmates. Gigantic cesspools were in close relation to the water supply and every arrangement was made for the pollution of the air by regurgitation of gases from the water closets.

Quoting another article in the *Sanitary Record*, it suggested that the school had been both secretive and slow to react.[40] This was all very damaging to the school's reputation – and was made worse when the *Lancet* returned to the attack a week later, reporting the recent RSA meeting:

> A resolution was unanimously carried that serious blame attaches to the masters ... The medical officer of the school was also blamed for not investigating the causes of the disease and not reporting the fact of the outbreak to the Sanitary Inspector. The Board likewise censured [Dr Bell] for not attending a meeting of medical practitioners in the town, convened by the Medical Officer of Health for the purpose of obtaining every information on the cause and history of the outbreak ... we should like to know if his reticence was due to pressure put on him by school authorities.

It also challenged Thring's fitness to continue as headmaster, for allowing healthy boys to visit infected houses: 'This fact the committee consider should be laid before the trustees of the school, and parents of all boys who had suffered ...'. It concluded: 'The Sanitary Authority have acted with spirit and determination and the censures they have administered will be a useful warning to other offenders.'[41]

Similarly critical articles appeared on that day in two Liverpool papers, right in Uppingham's recruiting heartland. The *Liverpool Post* believed that 'the commonest precautions have been recklessly disregarded',[42] while the *Liverpool Daily News* alleged that:

... letters and telegrams sent by anxious parents had remained almost unanswered ... Mothers, who fled in an agony of apprehension to Uppingham, had the greatest difficulty in obtaining access to their sick children.

It added, for good measure:

Perhaps by this time the logic of facts has convinced [that] even the autocratic will of the headmaster of an English public school is inefficient against the laws of nature, that sewage gas will bring enteric fever, however sternly he may set his face against it ...

The paper went on to describe the recent rise of the school, and its strong reputation for the teaching of Latin and Greek. But:

Perhaps when the cesspools are all cleared out and the water supply is beyond suspicion, and the boys are all back at Uppingham, the Local Government Board ... will send a teacher of elementary physiology into Rutlandshire. It would be a good investment of time on the part of both masters and boys, even if the [study of classics] were intermitted for a month or two.[43]

It says much for Thring that, although under such fierce attack, he did not surrender hope altogether; even when one paper described him as 'a bigoted old-fashioned hater of pure air and water'.[44] He replied to a supportive letter from Dr Jex Blake, his opposite number at Rugby:[45] 'I prize your telegram and letter exceedingly. It is very cheering to me in these heavy days to have a little sunlight let in.'[46]

School and town continued to conduct a smouldering dispute in the press in the days which followed, this time over whether it would be safe for the school to reassemble in January. Haviland, true to his role as MOH by putting risk to individuals ahead of risk to the school from further disruption, urged caution and the *Lancet* supported his view in an editorial a week later.[47] This contrasted with Tarbotton, who had completed his survey of the houses commissioned by the school, and who wrote more reassuringly to the housemasters: the latest analysis of the town springs by a leading London water expert, Edward Frankland, was 'most satisfactory'.[48] Frankland also wrote to Christian himself, recommending that the wells in the garden of his house be linked to the house by lead pipes.[49]

By December the worst from the press was past. *The Times*, much more sympathetic to the school this time, commented critically on the RSA's decision over the publication of the various reports awaited. Its intention to make them public as soon as they were completed, but before they had gone to the school trustees, would, *The Times* said, be 'a partial and premature act'.[50] Barnard Smith defended himself in a memorandum disputing charges in the *Medical Examiner* that the RSA was actively antagonistic to the school.[51]

It remained to be seen how much damage the flurry of publicity had dealt the school. What the experts' reports would eventually bring was hard to predict, for the signs were contradictory. Thring was cheered by a rumour that one inspector from London had been forthright: the RSA had much to do in the town and 'If work was not done, and done quickly, they (the LGB) would send down their own men and engineers, and charge it to the parish'.[52] However, one document in the LGB files gives a bleaker indication of what was to come – a copy of a letter written by a Mr Compton Maul. Maul was the parent of a boy in Thring's own house; he had written the letter to Thring on Christmas day 1875. In it he described how he had recently run into Haviland in the street in Northampton. They had talked for about ten minutes:

> Mr Haviland did not say that it would be a year before the school could possibly re-assemble, and that the locality would never be healthy again ... the purport of what he said was that it would be a long time before the boys could return with safety, that the sanitary condition of the school was very bad, that the privies at the school house were in a disgraceful state, that the boys did not get enough to eat and drink, that those who paid 6 guineas a year for extra meat did not get an equivalent for their money – in fact he said quite enough to deter any father from sending his son to Uppingham. I beg you not to suppose for a moment that I acquiesce in what Mr Haviland said – the illness at the school is a great misfortune and I fancy will deter many from returning.
>
> P.S. My boys are anxious to return to Uppingham and I shall be glad to send them there, provided I can be assured that the place is <u>safe</u>![53]

On discovering that Haviland was releasing the findings of his report ahead of publication and to the detriment of the school, Thring forwarded a copy of this letter to the Board, with the observation that he had only ever met Maul once in his life. He added: 'It is hard having Mr Haviland as our judge. Money has not been spared since 22 years ago I began life at Uppingham, with 25 boys in the school and but one boarding house.' He also enclosed copies of yet more exchanges between himself and Haviland, adding: 'I venture to think that the course pursued by the medical officer and the views entertained by him of his duty are not such as would be approved by the Board'.[54] In addition he requested that all the relevant background documents to this complaint be laid before Sclater Booth himself. An LGB minute records that 'the papers are with Mr Rawlinson, and he cannot spare them today'. It is likely that the LGB's officials were watching their files fill up with a sense of foreboding. They had insufficient enforcement powers and too few civil servants – and there were 700 other authorities with which they had to deal.

Thus 1876 began with town and school in a state of armed truce, both keenly awaiting the publication of the various reports: four in all. Alfred Tarbotton, commissioned by Thring and the masters, was to report on necessary improvements to the houses. A privately-sponsored survey by a Nottingham engineer would not carry the same weight as an inspection by government experts from London, but it would at least enable the school to show it was not complacent, and perhaps to put improvements in place which would help to speed up the return of the pupils. Second, the RSA's action in seeking advice from Rogers Field had brought into the dispute one of the most expert drainage engineers of his day.[55] He would advise it on the state of the town and on any necessary improvements to its streets, sanitation or water supplies. Third there would be the LGB report from its chief engineering inspector, Robert (later Sir Robert) Rawlinson. Finally, both town and school had appealed to Haviland for advice, and both had strong reasons for hoping that he would support their view of events. The stakes were particularly high for the school in this respect; parents would pay particular attention to Haviland's views, given their demands in October for him to be called in.

Tarbotton reported quickly. His inspection was confined to the boarding houses rather than the school classrooms and other buildings; it was sent to Thring and the trustees some time before Christmas 1875. Rawlinson also received a copy. Conceding that all the twelve houses had been defective in drainage arrangements in various detailed ways (drain layout, faulty joints, poor ventilation and inadequate flushing), Tarbotton judged that the defects were only those 'too often found in the most modern houses and mansions'.[56] He did not spare the RSA, pointing out that the four 'country' houses (up on the hill some way south of the town: those of Bagshawe, Earle, Christian and Rawnsley) had no possibility of connecting to a public sewer unless the system was radically extended. Prospects were not much better, however, for the 'town' houses, because the sewer there was 'deficient in depth and construction and totally unventilated'. At least one house (Rowe's, in the High Street) had had to construct a cesspit because the local authority had specifically banned it from connecting up to the sewer, presumably for fear of overloading the system. Tarbotton conceded that the Lower School was, despite its comparative newness, very defective.[57] It had unventilated and sealed-up cesspools, and some leaky drains. He concluded, however, with the statement that very extensive works had recently taken place, and that the Lower School now had a new well for its water supply. He reported that the masters had all been very cooperative: 'Works of sanitary improvement have been and are already being carried out in every house' (which he listed). Finally he urged the RSA to seek a 'better source of [water] supply unless private enterprise be more active'. He also drew up a memorandum of guidance for present and future engineering improvements, ventilation and flushing. He made no mention of miasmatic issues, and any contagion risks lay outside his brief or area of expertise.

If Thring had hoped that the positive tone of this report would persuade the trustees to agree to the school's reopening, he was quickly disappointed. They met again on 28 December and decided that they could make no decision until they had seen Rawlinson's report;[58] they were particularly keen to see what the judgement from London would be. Thring could not know as yet how long Rawlinson would take to produce his verdict, but by mid-January he believed that he had an assurance from the LGB that Rawlinson would say nothing which would prevent the reopening of the school just before the end of January.[59]

Rogers Field's report (commissioned by the RSA) came out on 6 January. It was carefully researched, wide-ranging in its scope and filled with technical detail. After describing the growth of the town and the sewerage improvements made in previous years, he detailed the sanitary arrangements in all the 379 properties in the town which the RSA's inspector had visited at his request.[60] Although brought in to advise the RSA, Field was forthright about the poor sanitary state of the town. He reported that the south sewer – just three years old – had been well constructed and was in good condition, but its seven ventilators were all blocked. Older sewers in the central part of the town had faulty joints and were poorly ventilated; many of their ventilators were choked with dirt. In the eastern part of the High Street, gradients were very flat,[61] so that cellars from some properties could not be drained into them and thus remained flooded. None of the sewers had any provisions for flushing; some sewage ran out of storm outflows and too much of it backed up in manholes because of inadequate gradients. However, Field also noted that although a system of sewerage had been constructed, it appeared to be very little used, 'the greater portion of the town still draining into cesspools, many of which are very badly situated and offensive'. At the sewage farm in Seaton Lane, he noted some defects, and suspected that tanks were too small and had been emptied only very irregularly – although he noted a recent change: 'Now they are emptied every fortnight'. Finally he noted that there was no public water supply, and that many of the private wells appeared to be contaminated. All these facts led him to conclude that the cesspits must be abolished, and that the water-carriage system of sewage disposal should be completed, rather than a dry-earth method being introduced. He believed that it would be better to lay new, deeper sewers than to attempt to relay the existing ones where they had faults. He was emphatic that a mains water supply was needed, both for drinking purposes and to aid sewage disposal and drain flushing.

Field's report was not merely about the state of the town. Within the school itself, he had visited every house and he described matters in great detail. He found that many of the cesspools were unventilated or poorly so; some were too close to wells. There were examples of drains passing under houses, poorly sited and poorly maintained water closets, and a number of other engineering deficiencies in sinks, baths and lavatories. At least two houses had poor connections to the town sewers. Many wells were too near to cesspools and drains. But Field also laid great emphasis on miasmatic problems arising from gases and foul air – including poorly-ventilated soil pipes. He agreed with Tarbotton's recommendations, and he too praised the cooperative and positive attitude of masters who were carrying them out.[62]

Rawlinson's findings (for the LGB) followed a few days later, on 12 January. His report was much briefer than the others – perhaps reflecting his instinctive reluctance for the LGB to be drawn too far into disputes of this sort and his belief in the persuasionist approach. He concentrated on issues of water and drainage, explaining how Thring and the masters had requested an enquiry, how he had met them on 16 November, and that he had then joined Tarbotton and Field on a tour of inspection to see the work already being done to put the existing town sewers and drains into better order. He listed past complaints against the RSA and the works which it had undertaken since 1857, with the costs and the loans taken out to cover them. He added: 'As the school-houses have been increased in number, reiterated applications have been made by the masters to the local authority for public sewerage accommodation, and also for an improved water supply'. He concluded: 'During all these years and after all this expenditure of money, the main sewers have been practically useless, if not in some respects mischievous', owing to inadequate maintenance. The report then added a new dimension:

> There has been local opposition by the owners and ratepayers, and so the local authority has performed its duty only imperfectly; making sewers at considerable cost for parts of the district but never completing them, or even serving notices for the abolition of private cesspools and the execution of house drainage.

It condemned 'owners and ratepayers, by their own obstinacy in not draining their houses'. He also brought up the miasma issue: 'There is most unfortunately a strong prejudice in small rural towns and villages against sewer ventilation because, it is said, the openings permit bad smells to issue'. Rawlinson believed that once Tarbotton's recommendations had been carried out, 'the school will be in as complete and satisfactory a state as the best modern sanitary science can put them'. He approved of Field's inspection work.[63] He reiterated the dangers from contaminated wells. All in all he was highly supportive of the school.

Rawlinson's report in particular impressed the trustees. Six days later (18 January) they confidently agreed that the new term could begin on 28 January,[64] although Wales, who was both a trustee and a leading RSA member, wanted the decision to be delayed until the one remaining report – Haviland's – had been published. It is possible that his dual role had given him a more detailed preview of Haviland's thinking than the other trustees could have had.

Notwithstanding Wales's opposition, the trustees instructed Thring to inform the parents that all the necessary sanitary measures had been carried out, and that Dr Childs had 'been appointed science master and charged with all sanitary arrangements in connection with the school'. Thring wrote to his brother that 'we have had a squeak for it',[65] and that Birley and Jacob had had some difficulty in getting the motion passed at the meeting. Again he complained bitterly at how little the town had done, either in recent weeks or during the previous five years, and reflected gloomily on the likely impact on the school's entry lists of all the bad publicity. He made it clear that he would have resigned if the motion to start the new term had not been passed:

> It would be ludicrous, if it was not so important, to see my trustees ... sitting in sol-
> emn conclave playing with other men's lives ... Yet there they are, totally ignorant
> of the business of the school, also passing judgement on us and our work and our
> fortunes ...[66]

Haviland, having originally been called in by both town and school, albeit for very different reasons, had made a number of visits to Uppingham between October and Christmas 1875. He had continued to write in both the *Lancet* and the *Liverpool Daily News* that as he had no evidence that the proposed structural improvements to the houses had been made, he could not recommend the return of the boys.[67] He had taken his time over his report, encouraging all parents whose sons had been affected by the disease to contact him.[68] As MOH he was far more concerned with the proper responsibilities of his role towards public health and safety than with the adverse impact on the school of his report being published just before the planned date for the pupils' return.

Thring had some premonition of what Haviland's report might contain, thanks to the letter which he had received just before the end of December from Compton Maul. In the week before the new term was due to begin, he confided to his diary: 'Private copies of the indictment of the school going about. The masters are very troubled, and there is plenty of reason for it, for it is ... clever and scurrilous to the last degree'.[69] Three days later he added:

> I have heard from London that the report is going about there. Beale (a medical spe-
> cialist, supportive of the school) is very disgusted. Jacob and Birley have also seen
> it. I hear the Bishop of Peterborough says the trustees must notice it. This is a fresh
> danger ...[70]

Haviland's report ran to fifty foolscap sides – by far the longest of the four reports. It opened with a graphic description of some of the likely preconditions for a typhoid epidemic, including 'the domestic pig, when fed on the offal and blood from the butcher's slaughter-house'. It emphasized the paramount need for speedy investigation of the first case in any outbreak, and the advisability of keeping young people away from any infected house. The history of the outbreak was then given chronologically – from the death of Hawke in June at the Lower

School, via Chapman's explosive visit to its underground chambers, to the clutch of cases in October which had caused the four deaths. Haviland stated that there had been thirty cases by 12 October and that by that date not 'a single step had been taken towards investigating the cause of this lamentable outbreak'.[71] He cast doubt on Thring's claim to Liverpool parents on 11 October that they did not need to contact Haviland because the school had already summoned him, suggesting that the telegram had not gone off until the evening of that day, and was thus sent only after parental pressure had started to become overwhelming. Mrs Richardson's complaint that the true state of her son's case had been kept from her until it was too late 'made a deep impression on me ... I found indeed, that she had reason to complain, and moreover I made the further discovery that she did not stand alone in her unmerited trouble'. He criticized Hodgkinson for allowing his cook to go home to Caldecott (a village five miles to the south), probably causing the death there of an eighteen-year-old maid who lived in the house next door. Dr Bell was censured for failing to attend the meeting of town doctors called by Haviland on 20 October – in contrast to Dr Walford, whom Haviland praised for attending it despite 'serious illness'. He implied that Thring had put pressure on Bell to stay away. Bell's subsequent complaint to the LGB (after Haviland visited one of his patients) was criticized at length.[72] Haviland rejected the charge, made in a petition to the LGB from John Hawthorn and twenty-two other town residents, about his 'having made various unofficial statements'. He was particularly incensed that the infected houses had not been closed to other boys at an early stage: this might reflect at least a belief on his part in the risks of contagion as well as germ theory, although not conclusively so. He confirmed his conviction that the epidemic originated in the Lower School, citing Professor Wancklyn, a water analysis expert, who had examined water from its wells[73] (although the professor was later to claim that his views had been misrepresented). He also believed that Nash, from far-off Redgate, had contracted the disease by swimming in infected water. This section of the report described the course taken by the stream flowing out of Hodgkinson's garden, and painted a graphic picture of its progress through the town:

> Above this point it is pure ... It [then] flows along the south of the town, receiving first an effluent which may be traced to the overflow of the town spring. It is contaminated by the oozings from the site of the old gas works ... after this it receives the drainings from manure heaps, of a cowshed, a pigstye, a stable, and other accumulation of filth ... and before passing under the bridge it becomes still further polluted by the small overflow of a cesspit in the neighbourhood of the national school. After this the stream skirts the cemetery, and receives its drainage; it then flows on beyond the town and becomes the feeder of the bathing place and swimming pond! There the water becomes so filthy, that from the information of men who were old Uppingham boys, many wisely declined to enter its befouled waters ... The poor Nash had bathed

in this filthy pond as late as the 14th September, if not later ... This brook then passes to the south of Bisbrook, where I am informed it is used for brewing purposes.[74]

There then followed four pages detailing shortcomings in the Lower School's water-closets, drains and cesspits, before he turned to the shared drainage between Mullins's boarding house and Paul David's private dwelling close by. He believed that when water was discharged down one drain, it resulted in foul air being forced up the other. Thus he introduced a miasma speculation. He also criticized the fact that, although neither of the two wells at Mullins's house was fit for drinking, boys had somehow been able to use them in direct contravention of their housemaster's instructions. These wells should have been used only for washing, as their water came from a tap in the WC cistern supplied from a contaminated well.[75] He then referred much more briefly to deficiencies in the other houses.[76] Overall his tone was in marked contrast to the inspection report of 1868 for the Taunton Commission, which had praised the school's domestic arrangements as 'perfect'.[77] That report had been written some years earlier, of course – and at a time when there was no crisis hanging over it. In the intervening period, public expectations about sanitation had risen markedly; numbers in the boarding houses had increased, and issues arising from wear and tear and the passing of the years may not have been fully addressed: we cannot be sure.

Haviland then moved on to reporting on the sanatorium. He had visited it with Field and had found major deficiencies in the cesspits. He criticized 'a great want of nurses ... the matron complained that all authority over them was denied her' (the reason cited for her resignation), and he condemned the practice of waking boys up for regular feeding when what they really needed was sleep. He was especially critical of Bell's treatment of one patient there, John Millington Sing from the Lower School.[78] According to Haviland, Bell had advised that Sing be fed at thirty-minute intervals without fail. However the nurses had found that it took quarter of an hour to wake the exhausted boy. By the time he was fed and had gone back to sleep according to Bell's schedule, it was time to wake him again. Haviland had happened to call in at the sanatorium just before leaving Uppingham and had sensibly said sleep was the one thing needed. 'The advice had been followed, and the boy slept soundly for several hours and eventually recovered. I could not see a boy struggling for life, and not give him help in the time of his need in the form of advice, which I knew to be sound.' Dr Bell, however, had subsequently written to Haviland to complain that he had interfered with his order.

Finally, Haviland criticized a number of other aspects of the school regime. He returned to issues of miasma and ventilation: the studies and dormitories were too small and/or overcrowded: there was allegedly less cubic space per pupil than that provided for prisoners in the Daventry lock-up. Many studies were ill-ventilated or not ventilated at all. 'It is absurd to suppose that a boy can study in an unventilated box, without a fireplace, and heated in the winter by a hot air pipe.' He implied that the food was too economical and insufficiently nutritious. He criticized the lateness of breakfast (caused by the timing of early-morning lessons) as a possible promoter of disease. Haviland also suggested a wide range of possible explanations for the outbreak – including poor diet, over-demanding school routines causing weakened resistance, swimming in foul water and/or drinking it, and poor drainage. There was no discussion of the major disease theory questions as such: suggestions of both infection and contagion are interspersed with frequent references to poor ventilation and sewer gases. There are also theories as to how 'the poison is generated in the excreta of an affected person after they are voided'. Haviland believed in the existence of a 'process of putrefactive fermentation which they invariably undergo when massed in cesspits etc', and asserted that 'the poison is liable to gain access either to the air or the water' – suggesting that he was at pains not to rule out miasma causes, particularly in the case of Kettlewell from the Lower School, who (he believed) had contracted the illness 'by being exposed to the influence of sewer-gases, emanating from the unventilated cesspool' there.[79] Haviland had also considered possible sources of contaminated drinking material but, while he recognized 'the need for a further and purer water supply', he believed that this was less pressing than dealing with the cesspits. On the other hand 'only by such a means can you guard against the present and future influence of the disease ...'. He had pondered – and rejected – the idea that milk from Mr Wortley's[80] cows at the nearby village of Ridlington might be to blame.

Haviland reused a number of the statistics which had appeared in his report on the combined districts eighteen months earlier, but it is noticeable that he drew rather more optimistic conclusions about the general state of health in the town than in that earlier report. He also asserted that at the recent doctors' meeting which had become notorious for Bell's absence, the other two GPs (Walford and Brown) had claimed to have had only three typhoid cases between them in the town in the previous two years – and those only in one property. Annotations and names listed by Thring on the copy of the report still held in the school archives suggest that Bell disputed this.[81] In the final paragraph of the report, Haviland thanked the members of the RSA for their support 'throughout this tedious investigation' and expressed the hope that 'whatever course you decide to adopt will be successful in securing one of the healthiest and finest sites in the country from preventable disease'. He even implied that it was the RSA alone which had originally invited him to intervene back in October, whilst omitting the fact that the school had made a similar request a day or two earlier.

To the school, Barnard Smith now added insult to injury. In distributing Haviland's report, he attached a long memorandum setting out his version of the events which had led up to it. He too emphasized the lack of cooperation from Dr Bell; he believed that, in complaining about Haviland's actions, Hawthorn and his fellow petitioners had failed to produce evidence or witnesses to support their complaint against Haviland's conduct.[82]

Hodgkinson felt bound to respond individually to criticism of his conduct. In a short pamphlet dated 10 February he defended his actions, 'pleading guilty to ignorance of the origin and nature of typhoid', having [he said] never experienced a case in twenty-one years. He disputed points of detail in the cases of both the Southampton pageboy, and the Caldecott maid; things were not as Haviland had portrayed them. The Chapman gas explosion had been greatly exaggerated, and the cesspit system around the town was one 'which the local authority did not raise their little finger to alter or improve' (as Rawlinson had pointed out): 'The blame of course rests primarily with the Sanitary Board'.[83] Hodgkinson sent a copy to Wales, who responded to an old friend with uncharacteristic gentleness. The distress which Haviland's report had caused was understood (Wales said), but the RSA had felt compelled to publish it, or – like Thring – it would have been accused of secrecy. He was at pains to explain that Haviland's was an independent voice; that no one was blaming Hodgkinson personally for the state of the Lower School cesspits, but that the first case of illness there should have led to an investigation. The RSA really was doing all it could to remedy problems in the town, but it was inevitable that ventilators would periodically become blocked. He also explained that the LGB was being very slow to grant the RSA increased powers through new bye-laws.[84] Hodgkinson wrote back on 25 February,[85] only a little mollified. Wales replied again on the 29th,[86] but it was a dialogue of the deaf.

Haviland owed the school no loyalty, and he had no reason to protect it. It was his responsibility to try to act in the interests of the community as a whole. Even so, it is hard to assess his thinking in producing a report which was so much more critical of the school than those of the other three experts, and which went well beyond dealing with issues relating to the epidemic. He took a highly hostile view of the school's past actions, and ignored any reputational difficulties which it would face through his criticisms, and/or if the disease returned. He must surely have had some inkling of the damage that such a wide-ranging report would do. It is not clear why he turned so decisively against the school after his early, comparatively civil meetings with Thring, or why he appears subsequently to have been determined to show it in the worst possible light. He may well have been incensed by what he perceived to be Thring's high-handedness in repeatedly seeking help from the LGB. It was inevitable that Thring would feel that Haviland had concentrated exclusively on the immediate reasons for the outbreak, while saying little about the RSA's inactivity in the past. Haviland had been outraged by the complaints to the LGB by Dr Bell, whom he singled out for particular criticism. He had a low opinion of Bell's skills as a physician; in seeking to apportion blame as part of his enquiry, his anger focused on a man whose work he could assess with some expertise, even if he mixed this with a certain amount of hindsight. In this way the school seems to have become the displacement focus for Haviland's anger about Bell.

Haviland also showed little regard for the day-to-day realities of housemasters' lives. With autumn drawing on in October and the hours of darkness increasing, the morale of both pupils and staff would have been increasingly tested; with long evenings in candlelight there would surely have come a greater sense of fear and foreboding if they were unduly confined to their houses. He allowed little for the fact that, while it was fine in theory to make rules for boys about where they might go and what they might drink, enforcement was difficult in practice. Housemasters taught classes and could not watch their boys all the time – nor would it have been practicable to guarantee to restrict pupils' movements and to separate them from their friends in other houses for long periods in order to keep them out of harm's way. Many lessons took place in house dining halls, and boys needed to move around the town to attend them. His criticisms of the broader aspects of school and boarding house management smack of someone pursuing a personal as well as a professional campaign, with (it could be argued) a selective use of facts. Yet he regarded the school's buildings and day-to-day regime as fundamentally unhealthy in many aspects, however disease was spread, and this was an opportunity to correct them. Although he made some criticisms of the state of the town – notably at the national school cesspits – these were moderate in comparison to the evidence already cited by both Field and Rawlinson. They contrasted starkly in their moderation with the blame which he heaped upon the school. Perhaps it was the fact that there had been nearly forty cases in the school but less than half that figure in the town[87] which convinced him that the school should bear nearly all the blame. He probably took into account that the town had eight times as many people, and that there was no evidence of any epidemic amongst those of school age except in the school itself.

Ultimately it can be argued that Haviland hedged his bets over the root cause of the epidemic. While he concentrated on problems of infected water, his bombastic style reveals something of a scatter-gun approach of criticism, partly based on miasmic and partly on other theories. It adopted a multi-causal approach to the epidemiological issues, vehemently condemning dirt of all sorts and describing vividly certain areas of town and school – but refraining from relating them specifically to the various disease theories, and from speculating as to precisely how these areas of risk might have caused pupils to contract the disease. It is hard to avoid the impression that his zeal for public health went far beyond his knowledge – but he probably felt that *all* the contemporary theories about the cause of epidemic disease might have validity, or that the causes of typhoid – dirt, foul water and negligence – were too obvious even to be worth discussion. Both Rawlinson and Field had raised miasma issues, too (but neither of them had a brief which would have caused them to discuss contagion). If Haviland equivocated over precise causes, he did so because there was no conclusive evidence about the source of the epidemic, nor of any carrier, and because of the limited state of knowledge at the time.[88]

Just over a decade after the publication of Haviland's report on Uppingham, nearly all his health-specific and wider recommendations appeared in the first edition of a book by the MO at Rugby. Over the next twenty years, Clement Dukes's book *Health at School* would become a much-reprinted handbook for the running of a good boarding establishment.[89] The wording of the introduction to its first edition, together with the brevity of its bibliography, suggests that very little literature was available before that date which would have given Thring and Bell any precise guidelines about how to handle a school epidemic.[90]

Haviland would make only occasional appearances in Uppingham over the course of the next year – usually to advise either on further cases of small-scale illness or on proposals to prevent them. His work was largely done at this point. He did not create the antipathy between Thring and Barnard Smith's RSA, but he certainly sustained it. If he had developed a strong personal dislike of Thring during his meetings that autumn, it is unlikely that he was wholly to blame – the two men were temperamentally similar in some ways. But the way in which Haviland promoted his public health crusade ensured that any remaining spirit of compromise between the two sides rapidly disappeared. .It also created a lasting and bitter enmity with Dr Bell – one which Bell would pursue relentlessly through the year to come. In that sense, Haviland was the catalyst for the events which lay ahead.

6 SPRING 1876

'A terribly cold north-east wind and a slight fall of snow, looking threateningly for more ... it seems still very uncertain when the school return, maybe 21st (but probably not), or 28th or 4th February', wrote Mrs Hodgkinson from the Lower School to her daughter just after New Year.[1] With the trustees' meeting a fortnight later came the decision that it would be on 28 January: 'Pray God keep us this term', wrote Thring in his diary: 'Masters' meeting this morning. Had to speak to them strongly about tittle-tattle.'[2] For a passionate and sometimes excitable man, he was surprisingly at peace at this time – outwardly at least. At Christmas he had bidden farewell to one of his most stalwart housemasters, Theophilus Rowe, departing to become headmaster of Tonbridge – a man whose organizational skills, diverse talents and encyclopaedic knowledge on many topics would be greatly missed. But after the busy and frustrating weeks of presiding over an empty school, Thring could now get back to proper work. He had received Haviland's report just before term began. In between bouts of raging at its perceived injustice, he seemed almost resigned:

> As we have often said in old days, 'If this thing is of God, it will stand; if not, let it go ... It illustrates what I have so often pointed out – the impossibility of getting at the truth in a complicated matter ... I was almost amused when I read it, at the ease with which I was made out a liar and a scoundrel ... It marvellously opens a man's eyes when he has once or twice seen himself pictured in the 'devil's looking-glass'; he gets a sounder idea of man's praise and blame, the latter especially. I may yet go down to posterity as the great flogger, and 'bigoted old hater of pure air and water', and senseless, unfeeling tyrant over boys which these fellows paint me.[3]

He cannot have been pleased to receive a letter from Dr Bell asking whether it was he who had suggested to Haviland that Bell had been slow to recognize the nature of the illness[4] – or by another lengthy and critical report which appeared in the *Lancet* that day. It concluded that:

> The school authorities assumed a grave responsibility ... Sad as the lesson is, it will not be without value if it teaches [them] the necessity of trusting less to their own omniscience and more to the guidance of those who, by special training, are best qualified to give advice and act in such emergencies.[5]

An editorial added to the annoyance, by backing Haviland in every respect
– including his comments about the wider management of the school. A few
days later, however, he felt that school life was gradually returning to normal.
'There are thirty new boys ... and 305 on the school-books, so we have not suffered an appreciable check',[6] he wrote cheerfully on 29 January, although he may
have been in denial about the real state of pupil numbers. The *USR* suggests that
over fifty pupils under the age of eighteen left the school in October or December 1875, while only just over thirty joined in January 1876. He could not yet
know whether things might get worse; in the event, another sixteen would go in
March, of whom some can be identified as transferring to rival schools such as
Rugby, Repton and Clifton. Not only that, but eighteen of those who left were
from the north-west and another thirteen from London – suggesting a parental grapevine of gossip adversely affecting the school. Mullins's and Christian's
houses seem to have been especially hard hit, with eight losses each – hardly
surprising, considering the large number of typhoid cases in the former's house
during the October outbreak (and the death of his son), and the fact that the
late Stephen Nash had been one of Christian's pupils at Redgate. Rowe's house
was also affected by diminished numbers, but it was not unusual at that time for
masters to take boys with them when they left a school, and several of them had
gone with him to Tonbridge.[7] It was just as well that Uppingham's numbers had
been allowed to creep up above Thring's optimum 300 in the previous few years;
at least it allowed for a little unnatural wastage now.[8]

As the term gained momentum he began to rue the large number of abnormal
matters still to attend to; he complained that he had little time for intellectual
work and teaching. But he was also cheered by the progress being made on the
trial borings: 'The water works on the hill are going well',[9] and the fact that the
parliamentary processes were under way.[10] The water bill got its second reading
in the House of Commons on 25 February.[11] The rebuttal in the *BMJ* by Professor Wancklyn of Haviland's assertion that Wancklyn believed the Lower School
water supply to be the source of the typhoid outbreak, was further good news
for Thring.[12] The RSA seemed increasingly uncertain about the best attitude to
take over the water supply question. It did not wish to be seen to obstruct the
public good, but it was hardly likely to be enthusiastic about endorsing a private
company outside its long-term control, and it was also determined not to let
Thring seize the initiative. On 9 February it instructed Rogers Field to 'ascertain
the best and most economical mode of providing a water supply'. In order to protect its position, it also instructed Brown, its clerk (at a meeting on 11 March),
to oppose Thring's bill for a private company, or at the very least to get a clause
protecting its interests.[13]

February arrived, and with it came a tone of greater cheerfulness in Thring's
diary: 'The first week over, such a blessing, and time, the great healer, moving

slowly on, carrying us, please God, out of immediate danger by degrees'.[14] He did however express doubts to Skrine, one of his closest confidants amongst the masters, on an afternoon walk. Skrine recounted later how, after revealing some plans which Thring was making for the further improvement of the school, 'he (Thring) dropped his voice to add: "If we are allowed to go on working together"'.[15] The fears were well founded. On 20 February Thring's diary records:

> This morning I have entered once again the valley of the shadow of death, and the dark creeping blackness is coming over us again. Cobb (housemaster of one of the smaller houses in the High Street – one not previously infected) came in to see me before chapel to tell me he was almost sure that he had a case of typhoid in his house. Poor fellow! He quite broke down as he told me ... The town has ... neither flushed the drains nor disinfected them, done nothing except the ventilators they were compelled to put in ...[16]

And so the roller-coaster of hope and despair began all over again. Lessons had been learned from the previous outbreak; this time notices were immediately sent to all parents to inform them of the exact state of affairs. All the boys in Cobb's house were sent home, and Thring braced himself for a rapid exodus of many from other houses. It was not quite as bad as that, thanks partly to the fact that a suspected case in Haslam's house nearby subsequently proved negative. The scare prompted a visit of two other housemasters, Bagshawe and Christian, to the RSA to demand action, and they reported that the guardians 'seemed frightened at the gathering storm'. Their demand seems to have been successful too, for Thring noted on 22 February: 'For the first time today the sewers have been examined and found foul enough to account for any fever. The rector was hauled to see it, and has heard some plain truths too, I understand.'[17]

The Uppingham parents were 'wonderfully steady'[18] at this moment; only one wrote critically. Jacob and a number of Liverpool families sent a demand to Sclater Booth for urgent action from the LGB. Tarbotton reported that he had been back with a medical expert to check that all was well with the house improvements. Thring spent the whole of 28 February closeted with Jacob and Birley (whose son was a new boy that term) while he wrote a new memorandum to the LGB; these two would still support him even if other trustees kept their distance. Dr Bell meanwhile was trying to assure at least one parent (a Mr Dalison) that his son did not have typhoid – although, typically, he was also busying himself in writing to the father of John Millington Sing, asking for help in refuting some of Haviland's recent accusations about him.[19]

The *Lancet* now reminded its readers of its earlier doubts about whether the boys should have returned, and repeated Haviland's warnings about premature reassembly.[20] Its cautious attitude was supported by suspected new cases in both Mullins's and Christian's houses during that week, although Dr Beale, called up rapidly from London, reported no cause for alarm, as there was no evidence that the fever was connected with the house itself. However, a few days later came news that one of Campbell's boys in Lorne House had been taken home by his parents, and had now developed typhoid symptoms. Thring wrote on 3 March that 'I feel quite sure this is the beginning of the end'. He expected that 'the school will slip away like a wreath of snow'. He fulminated against Wales: 'The chancellor's letters furnish us with an admirable barometer of what to expect from the powers that be in this place'.[21] The *Lancet* reported 'a case of enteric fever or two in the town itself'.[22]

It was inevitable now that anxiety – and further withdrawals – would gradually increase. A letter from Dr Bell on 4 March indicates that a number of boys had been to see him fearing (wrongly as it turned out) that they had the disease.[23] Telegrams began to arrive in large numbers from worried parents. A lengthy series of letters from Bell to other parents shows that he was again on the defensive, notably in his dealings with a Mr White of Bury St Edmunds, and with White's GP there, over a boy's diarrhoea symptoms and the circumstances in which he had been sent home.[24] There had been a complaint that Bell had not spotted signs of typhoid; parental doubts over his competence had not gone away. The reappearance of the disease, less than three months after the premature end of the autumn term and almost exactly four weeks after the return of the pupils, sheds little light on its likely source, given the length of time that carriers could continue to spread it. It could have been carried by a returning pupil or by someone from the town; it could also have continued to seep from drains to wells during that period.

If the LGB hoped that a new year would bring less activity from Uppingham, events soon proved otherwise, for another round of acrimonious disputes followed. In the first week of 1876 it received another petition from the town bookseller John Hawthorn, this time protesting against Haviland having leaked his enquiry findings ahead of publication. Hawthorn claimed that he could easily have collected many more signatures, and that people in the town hoped that the complaint would be vigorously pursued. Close on the heels of this petition arrived a copy of a resolution passed by the RSA condemning Hawthorn for his criticisms of Haviland, and stating:

> In the opinion of this committee, Mr Haviland, as Medical Officer of Health, has merely done his duty in investigating the cause of the late outbreak of typhoid; having perfect confidence in his integrity and ability they are prepared to sustain him in his course of action.[25]

The LGB also heard more from Bell. The new year was only two days old when he wrote to Sclater Booth demanding to see an advance copy of Haviland's report to discover whether, and if so how, he was criticized. At that stage the LGB itself had not seen a copy of the report because the RSA had not yet sent it one. Thring also contacted London: he too was keen to know what Haviland would say. The trustees were due to meet shortly and they would want all the relevant information to make a decision about recalling the school. Rawlinson advised the LGB that, since the school had implemented the proposals which both he and Tarbotton had recommended, nothing which Haviland might say about what had now become past history could materially affect this decision. In the circumstances, on 17 January the Board decided that the best thing to do was to dispatch to Thring a copy of what Rawlinson had written.[26] This action predictably outraged the RSA. It wrote on 19 January of its 'astonishment' that the Board had sent a copy of Rawlinson's report to the school without prior consultation and before Haviland's report was ready. This protest did have one positive outcome: an urgent journey to London by Barnard Smith himself to deliver a copy of what Haviland had written. But this visit in turn provoked extreme irritation at the LGB, where Rawlinson let fly in an internal memorandum with a fierce denunciation of what the town had failed to do in previous years, and of how little it had done in the months after typhoid first broke out. He contrasted this with the school's rapid action in hiring Tarbotton. Rawlinson added:

> It appears that the local authority and Mr Haviland think far too much about the school and far too little about the town, as it is clear that the school drainage was retarded by the defective state of the sewers. If these had been perfect, the Revd Mr Hodgkinson need not have constructed the cesspits he was advised were necessary.[27]

But the RSA remained unappeased. It wrote again on 3 February complaining that it had 'not received the courtesy and support which they [i.e. its members] might have expected from the Board, and without which their exertions, however strenuous and honest, are rendered in a great degree inoperative'.[28]

Things became no better for the LGB once Haviland's report was made public. Bell now disputed Haviland's charges point by point, in a long letter to Sclater Booth on 5 February:

> The whole report is open to very severe and just criticism: it contains much that it would have been well to omit, it quibbles over trifles, it enters so extensively into personalities in a manner so much to be regretted, it is so voluminous that the only important part – the cause of the outbreak – is almost lost, in fact I believed an efficient report would have been made in six pages instead of 46.[29]

The LGB decided not to send this letter on to Haviland, and merely stated that it could not take sides between the two doctors.[30] Mullins weighed in at much the same time, sending the LGB a copy of a letter of complaint which he had written to Haviland about the latter's comments on the size of his dormitories in West Deyne, and refuting his allegations that infected boys had been allowed to enter other houses.[31] The RSA then returned to the attack. It criticized Rawlinson for stating that his report had been drawn up in response to demands from the masters and the trustees. It claimed that the trustees had never been involved – a fact finally – and tardily – confirmed to the LGB by Guy, their clerk, as late as 25 March. According to the RSA, the masters had acted wrongly in asking it to take action, and then going behind its back to the LGB. It disputed Rawlinson's view that the school had completed its improvements, claiming that as late as 18 January nothing had been done at the sanatorium, not even the emptying of cess-pits. The RSA also asserted that it had been promised that Rawlinson's report would not be published before Haviland and Field had completed their work. It even criticized the way in which Rawlinson had gone about his investigation: 'He visited the town only once, and that for four hours. And this is called a royal commission!'[32] Rawlinson again told the LGB privately that the RSA was more concerned to protect its own reputation in respect of past events than to put things right.[33]

By now it was late February and news was filtering through of the new typhoid cases in the school. A new figure emerges at this point in the LGB's files; Mr Joseph Rayner wrote on behalf of the Liverpool parents, who had been notably more supportive of the school during this second outbreak than during the first one. He contrasted the activity of the school in recent weeks with the inactivity of the RSA, and demanded that the LGB exercise its rarely-used powers to order sewerage improvements in the town.[34] The LGB asked Thring for confirmation that the disease had reappeared. He replied on 28 February, in an explosion of anguish spread over no fewer than fifteen sides of paper. There was great alarm at the school (he said); Tarbotton had been called in again, but the town had done nothing; sewers remained unflushed and the wells were still dangerous.[35]

During the first fortnight of March Thring was in touch with the LGB three times about new cases – first in Mullins's house, then in Christian's and finally in his own on 13 March. Worse still, he had to report that the new well which the school had sunk outside the town for its own use had been pronounced unsafe by water experts.[36] All this had convinced him that there was no alternative to breaking up once again on 14 March. Just as he had done a few months earlier when the school had to close in November, he attempted to imply that it was the LGB's responsibility to act on behalf of the school: 'It is for the London authorities to determine what course of action should be taken that will enable the school to return with safety to Uppingham'.[37] He had little hope that the LGB would do so, for he was increasingly aware of its non-interventionist philosophy.

Meanwhile Haviland had been active. He reported to the RSA on how his return to Uppingham to investigate this second outbreak had been received by the school. The welcome had not been warm, and he hoped that the RSA would inform the LGB about it. He had gone there as soon as other commitments allowed, but meanwhile he had sent the inspector of nuisances to see Cobb, in whose house the latest problems had started:

> Mr Cobb being at school and engaged until 12 noon, Mr James called again at 12.10 and found Mr Cobb at home. He then delivered my message to which Mr Cobb replied that 'he would meet Mr Haviland either in the street or at the Falcon [Hotel] but he would not see him at his house.' I declined to meet him either in the street or at the Falcon and said I would only meet him at his house, where the enquiry must necessarily be made. Mr Cobb's reply to my message was: 'His compliments, and he had nothing to say'.[38]

The RSA went zealously into action on the strength of this account. A copy went straight to the LGB, which noted: 'It is most unfortunate that so much ill-feeling exists between the school and the sanitary authority, as it entirely prevents any cooperation between them in the present emergency'.[39] It then sent a further copy off to Thring, and waited to see the result. Thring replied almost at once. He claimed that Cobb's own account of the exchange was somewhat different. There must be some misunderstanding: Cobb had merely been informed that Haviland was back in the town, and Cobb had said that he had no reason to meet him. There had been no suggestion that Cobb would not speak to Haviland at the house. Thring was keen to defend his beleaguered housemaster, although he did concede that the misunderstanding was not helpful, and that in retrospect perhaps things could have been handled better. But he justified Cobb's actions by adding: 'But when we admitted Mr Haviland in October last to *all* our houses, he took advantage of it to make statements about our inner life'. He reiterated that Haviland had also visited their houses without their knowledge. He was keen to emphasize that the school had nothing to hide, and he had now given orders that Haviland might go wherever he chose. 'I wish in all things to show respect to authority', he wrote, and he expressed regret if any contrary impression had been given.[40]

Two days later Barnard Smith announced that he would be in London and would call on the LGB again that afternoon.[41] He assured it that he was very anxious to clear the way for 'immediate action', now that there were three new cases in the school. Judging by the LGB's notes, this was a more cordial and constructive meeting than their previous one. Encouraged by Barnard Smith's declaration of intent, and advised by Sir John Simon that evidence of the cause of the latest problems might be lost unless prompt action was taken – a view supported in an intervention by telegram from Sir Henry Thring – the LGB arranged for Dr Power from its medical inspectorate to be sent down to Uppingham. However, this visit had subsequently to be postponed when Power's mother became ill.[42]

Only a day later came another lengthy communication from the RSA. It had reverted to its earlier tone: it was wronged and misunderstood. It rebutted all recent criticism of its perceived inactivity and stated:

> The sanitary authority have met with antagonism where they had every right to expect co-operation, While they have been endeavouring to do all that existing circumstances allowed, they have been condemned as supine, indifferent and inactive.[43]

The reasons for this change of heart are not clear. The LGB eventually told the RSA on 6 April that it could make no useful comment on this polemic. The LGB emphasized its impartiality, and awaited the next tirade. Meanwhile it had been engaged with Brown over a host of routine matters. These included queries about whether the expenses of recent enquiries could be settled by post office order, details of a disputed surcharge as a result of a recent audit (which, keen not to inflame tempers any further, it eventually used special powers to remit) and advice on whether or not there would be a conflict of interest if the son of an RSA member were allowed to tender for the milk contract at the Uppingham workhouse.[44] It is unsurprising that the LGB was overworked, when it had to deal with such a diverse range of local minutiae.

With no sign of a respite in the recurring disputes between the school, the RSA, the LGB, Bell and Haviland, it was hard to see how the situation might be resolved and progress made. Yet for the school to do nothing at this point would only make its permanent closure more likely. It was time for some radical new thinking. This was unlikely to come from the LGB, which was also pursuing its policy of minimum intervention in local matters elsewhere. On 11 April *The Times* carried a report on Question Time in the House of Commons. In response to a question about the typhoid outbreak in Eagley, near Bolton, which had captured a number of recent headlines, Sclater Booth replied that the LGB had decided there would not be a public enquiry there. He justified this on the grounds that 'The sanitary authority seemed to be discharging their duty remarkably well, and they were engaged in prosecuting inquiries of their own through the agency of their medical officer and other skilled persons'. After its recent Uppingham experience, it is hardly surprising that the LGB had decided to continue its policy of non-interference in local affairs as far as possible.

The LGB had probably not grasped the full import of Thring's message on 7 March, in which he informed its officials that the school would be breaking up prematurely again, adding that: 'It is for the London authorities to determine what course of action should be taken that will enable the school to return with safety to Uppingham'.[45] A quite unexpected new debate, born out of desperation, had now begun to take place amongst the masters. Talk of migration to a place away from Uppingham altogether seems to have begun around 4 March. Credit is generally given to William Campbell of Lorne House, next to Mullins's in the High Street, and only a hundred yards from Cobb's. Campbell articulated this radical and adventurous possibility at a housemasters' meeting, with the memorable question: 'Don't you think we ought to flit?'[46] He was a man of long experience, not easily given to flights of fancy,[47] and the idea began to gain supporters. It was explained by Thring in a letter to his brother,[48] reflecting on how much more vulnerable Uppingham was in such a situation than better-known schools such as Marlborough and Winchester:

> We cannot hold the school together much longer; I doubt whether Tuesday next will see us with a third of the boys left here. They are melting away. The town will do practically nothing. This is ruin. We are thinking if we are deserted of migrating to the Lakes and getting all our classes together there till the summer, just to keep the school connection going.

He needed to seek the backing of those trustees who were likely to be supportive. Jacob and Birley agreed to meet him in Manchester to discuss the idea on 7 March. Birley told him that he had been contacted by the editor of one of the Manchester newspapers, who was being besieged by parents wishing to place adverts in their search for private tutors and alternative schools. But Birley also knew of a hotel keeper in Wales who was keen to get the school. This idea immediately caught Thring's imagination; he assured Birley that the housemasters were unanimous in supporting Campbell's idea for a temporary uprooting of the school along these lines. In saying this, he omitted to mention that there were doubters. Rawnsley, for one, was strongly opposed to another term being ended prematurely, and it must be likely that the innately cautious and conservative William Earle would have viewed even a temporary removal elsewhere with deep misgiving.[49] But Sam Haslam saw things differently. He wrote to a Norfolk parent (Mr Copeman) that migration was the only way for the school to survive: 'If we do not assemble somewhere while the work is being done, the school will surely vanish, or nearly so, the boys being absorbed by other schools and tutors'. He added: 'And among other incidental advantages, not the least is the good likely to accrue to every boy's character who shall come and share our difficulties in this crisis ... so we hope for the confidence and support of all parents'.[50]

Once back in Uppingham, Thring wrote to A. C. Johnson, the chairman of the trustees, on 8 March. Johnson had already been informed about the idea of moving; he had given Thring a free hand in principle on the question, but he now needed to know the details of what was proposed. He would have to persuade the other trustees, if he could. Thring reported to Johnson that he had returned from his visit to Birley, and:

> ... that the school will break up for its Easter holidays on Tuesday next, and that we shall reassemble in three weeks' time ... in some healthy locality away from Uppingham. Most probably Borth, near Aberystwyth ...[51]

Within a few days both local and national papers and the *BMJ* were report-ing that the plan would become reality.[52] In contrast to its rival the *BMJ*, which had deplored the fact that the school had been allowed to reassemble in January, before Haviland had given his specific sanction by declaring that everything possible had been done to prevent a recurrence of the epidemic,[53] the *Lancet* expressed sympathy for the school. But it did not mention Thring's plan to move. Its target for criticism on this occasion was Rawlinson, who, it said, had been:

> ... fully aware of Mr Haviland's doubts (about the school reassembling in January) and assumed a great responsibility in speaking so decidedly with respect to the sani-tary improvements ... The school authorities are therefore greatly to be pitied; they asked advice, and they spent their money freely in improvements, and now they have a second visitation which, we fear, must cause them severe pecuniary loss ...[54]

The trustees had yet to meet as a body to consider the issue. A few were very hostile as soon as they heard rumours of it – notably Wales. Thring wrote to Birley on 8 March: 'The rector has put his foot into it, having prevented a meet-ing of the trustees being called by saying there was no need. And he has already been using threats against us for our action. Let them do their worst.'[55] It lay within the trustees' constitutional powers to stop the plan, but some of them disagreed with Wales, realizing that to do so would almost certainly have meant the school's permanent closure. Thring recorded that in the days leading up to the meeting:

> The rector was sententious and threatening to one of the masters. 'The trustees would stop it all.' He might just as well try to stop a train with his finger. All the masters are unanimous. Legal or illegal, the only thing left is to do it in the best way. Change of air is the only possible prescription.[56]

A special general meeting of the trustees took place on 11 March – four days later than the masters had wished.[57] The minute book tells us very little:

> ... A statement of the Rev Edward Thring to the chairman in reference to the second outbreak of typhoid fever in the school was produced and read ... and that in conse-quence the trustees sanction the proposal of the headmaster to break it up ...[58]

Behind that bald statement lies a lot more dispute and acrimony. The decision 'that the masters be requested to furnish the trustees with a statement of the amount of space allowed to each boy in the dormitories and the provision for ventilation' suggests that the trustees wished to assert their authority, and that they would make no concessions to all the other problems the staff were facing at that moment. It also implied that the trustees were taking Haviland's criticisms very seriously. Such equivocal backing was unlikely to endear them to Thring and his staff. In the trustees' defence, it is clear that they resented being presented with what appeared to be a fait accompli – and the fact that Thring seemed to have briefed the newspapers before consulting them.[59] Both the *Stamford Mercury* and the *Manchester Critic* had carried a report the previous day suggesting that reassembly of the school was planned to take place 'in three weeks' time at some healthy locality away from Uppingham', and *The Times* had picked up the story that very day. The *Mercury* knew enough to tell its readers that the move was likely to be 'either to North Wales or the Lakes'.[60] The fact that Johnson stepped down from the chair with Sir John Fludyer deputizing for him for the second part of the meeting indicates that he was under criticism for exceeding his powers in encouraging the idea, and that as a body the trustees were seriously divided.[61] Whatever they decided now about the proposed plan, their position could clearly be only provisional, pending further developments and more information.[62] In the end they recorded no formal decision, but Thring's diary suggests a robust debate:

> The first battle of the new campaign fought today, and on the whole won. The trustees have sanctioned the break-up of the school, but on —'s [Wales's?] dictation would not put on record any expression with reference to the migration; in his own words, 'They knew nothing of the school till it came back again'.

They were, in effect, washing their hands of it, and Thring inveighed against one particular (unnamed) objector:

> He spoke of the ... buildings as burdensome to the trust, and endeavoured, whilst taking over some £14,000 worth of property from our hands, to saddle us with the burden of any occasional deficit on the small outlying debts ... Then he brought forward the day boys and the necessity of having a master here. I simply said I should not leave any one of my staff, but if necessary a man might be got to do it. But that they [the day boys] could come with us, and the trustees could pay a fair proportion of their board and lodging. Then he threatened that the trustees would have to cut down the masters' salaries. I quietly pointed out to Mr Finch, who was sitting next to me, that the scheme appointed that the tuition fees must first go to paying the masters.[63]

The *BMJ* and the *Lancet* both reported the premature end of term, the latter in terms much more supportive to the school than previously.[64] On the same day the RSA met and decided to press for a clause in the waterworks bill 'sufficiently to protect the rights and interests of this Authority: and on such a clause being obtained to withdraw the opposition to the bill'.[65] Maybe the implications for local trade if the school carried out its plan were becoming apparent to Barnard Smith and his colleagues.

Over the weekend which followed, Thring went through a variety of emotions, ranging from despair to elation:

A quiet day at last. Holy Communion, a very good sermon from Christian in the afternoon. When shall I spend a Sunday again as headmaster in this place? I had a feeling as I stood in chapel to-day, never – never; but then I looked up ... and then it came back to me again that much was left, and that even this place, with its deadly blight of dull, dead hearts worse than the typhoid, might breathe new life and remain a light ... One thing I feel more than I have ever felt, that a great shaping power is round about me, guiding, and ruling, and making, and moulding this fierce crucible work and fiery rush of evil and danger, and friendship and help all round about one ...'[66]

It would be surprising if Thring did not feel both angered and daunted by much of what he had heard at the trustees' meeting. Yet it also strengthened his resolve, for it was increasingly clear that he would have to decide his own destiny. He was buoyed up by the unanimous support of the masters; Earle had withdrawn his earlier doubts, and three others came to pledge their full support for whatever actions he decided to take. Skrine offered 'to put his salary into my hands'.[67] The LGB seemed to be taking a closer interest again. Jex-Blake (headmaster of Rugby) had been in touch to offer help; he would resist any temptation to capitalize on Uppingham's misfortune, and he would discourage parents from transferring their sons to Rugby. His support was timely but others were rallying too; *The Times* published a long letter the next day signed 'Pater Alumni', contrasting the 'plague-stricken city' and the supine attitude of the town with the resoluteness and imagination of the school in seeking a new location for itself.[68]

On 13 March, just as a typhoid case was confirmed in his Thring's own house,[69] he preached at the end of term service: 'Difficulties become tests of willingness and strength; all hardship, everything that tries life, when overcome, strengthens life'.[70] Skrine recorded that 'It was a day of wild winds and pitiless snows ... that afternoon we are gathered, with thin ranks, for the last time under our chapel roof. In a few hours we shall separate, to meet, who knows certainly where ....'[71] One omen seemed good. The Old Testament lesson (probably read by one of the masters) recounted the wanderings of Jacob in the wilderness. The choice of lesson was not mere theatricality on the Thring's part, although there were those who did wonder whether this story – with its eventually positive outcome – had been specially chosen. Skrine believed otherwise:

There came in deep tones from the lectern the story of an exile, who at evening lighted on a certain place, and heard in dreams the promise 'I am with thee and will keep thee in all places whither thou goest and will bring thee again into this land'. There were hands which turned the prayer-book's leaves to see if the ... lesson were the reader's choice, or (as it did) stood so appointed in the calendar ...'[72]

Thring wrote in his diary that evening that he felt: 'some strange ... good and marvellous divine purpose will come out of it all. Tomorrow I start for Liverpool, and on Tuesday for Borth and other places in North Wales. Borth seems likely to suit ....'[73]

As in the previous autumn the RSA had apparently taken very little initiative as these plans developed and no doubt gradually became known locally; little would have been secret in as tight-knit a community as Uppingham. Wales's opposition to them is unsurprising, given his personal coolness towards Thring, and the fact that, unlike many of his fellow trustees, he lived in the town; he would be able to appreciate better than most the likely economic impact of any prolonged absence by the school. It is also likely that he and Barnard Smith weighed up all the potential difficulties and risks already described which Thring's plan implied – including the logistical challenges, the risks he would face over the future recruitment and retention of pupils, the financial demands it would make on the masters and the prospect of uprooting all their families. Wales and Barnard Smith were no doubt resigned to the continuing wrath of the school, and probably prepared to ride it out. But where the ratepayers were concerned, in the absence of readily-available funds for wide-ranging improvements to drainage and water supply, the RSA was engaged in an act of brinkmanship. If it continued to do little it risked substantial criticism from the tradesmen if Thring carried out his plans; if it did too much, too soon, it would be lambasted just as fiercely for the rate rises which would follow. It is little wonder that its members were inactive for so long.

Thus in different ways both Thring and the RSA were damned if they acted decisively, but equally damned if they did not.[74] By mid-March opinion was stirring in the town, as the prospect of trade lost by the school's absence at last began to sink into the minds of local tradesmen. As a result, the RSA received a demand from ratepayers to be allowed to attend its next meeting. Brown (presumably acting on orders from Barnard Smith) attempted to head off the move by saying that the RSA was due to meet an LGB inspector at that meeting, so the idea should 'stand over'.[75] Undeterred, the same protesters sent a petition to the churchwardens, calling for a ratepayers' meeting to discuss what sanitary improvements the town might need.[76] The signatories represented a remarkable cross-section of local shopkeepers and suppliers of food, stationery and other services.[77]

The meeting was held on 23 March and appears to have been heated; rumours were spreading that the RSA might be planning its own water supply at a price which would undercut Thring's scheme. The prospect of low prices might be welcome, but what smacked of a spoiling operation against the school on the part of the RSA was not. At the meeting itself the opposition was voiced by supporters of the school – housemasters Candler and Mullins, together with Dr Bell. The shopkeepers and other tradesmen, having signed their petition of protest, appear to have played little active part – this time.[78] Four motions were passed: first that a private water company was preferable to one organized by the RSA; second that a surface supply would not do; third that the meeting disapproved of any spending by the RSA on plans for a surface supply, and finally that a copy of these resolutions be sent to the LGB. With great satisfaction, Bell sent just such a copy, on the following day.[79]

The RSA was put further on the defensive by a fresh bout of anonymous press correspondence in *The Times* over three days from 21 March.[80] 'A member of the school' wrote to reiterate how negligent the RSA had been. This was countered by 'One of the sanitary authority', who emphasized what a healthy place the town was (and had always been), and the sense of safety which its inhabitants felt. Revealingly, this writer also described the debt incurred through past sanitary improvements and the costs which present and future developments implied; it said much about his motives. Otherwise, the letter offered little new by way of argument, and many old charges were repeated. More conviction seems to appear in the letter carried by the *Manchester Critic* from 'One of the townsfolk' later in the week. It chided the paper for its pro-school stance, and blamed inadequate bye-laws and 'legal see-sawing' for delays in improvements. It pointed out that the speed of response from the Uppingham authorities had been rather more rapid than when the writer had himself lived in Manchester some years earlier, and, alarmingly for Thring, sought to sow the seeds of doubt as to whether Borth would turn out to be any safer a haven from disease than Uppingham had been.[81]

But for Thring the die was cast. Only one immediate hurdle remained: the attitude of the trustees again. They reassembled on 24 March in a mood of deep wariness. Motions were passed encouraging the RSA to carry out all Rogers Field's proposals, banning any housemaster from taking more than thirty boarders and stating that no boys who had been in a house where infectious illness had broken out should be allowed to return without the headmaster's permission. Even allowing for the dire threat to the school they governed, such decisions must have been intolerable for Thring, given that he had resisted their interference in management matters so determinedly over the years. The trustees dealt him a series of further blows. On being told formally by Thring that he had arranged for the school's removal to Borth, they resolved to put just £50 at his disposal to defray the costs of travel, board and lodging incurred by the day-boys who would be going with them. This was far from enough to cover such an outlay,[82] and only half the sum which they agreed at the same meeting should be given to their clerk for all the unusual extra work which he had recently had to do – a meeting at which they approved other grants for routine expenditure of over £1,000.[83] They declined to make any decision over whether to grant travel costs to the masters. They had decided to be trustees of the school *at Uppingham* in the most literal sense of the term, with no firm commitment to how they would react to future events. Maybe they reckoned that, as they controlled only about two-fifths of the school's total annual expenditure, Thring and the housemasters should look to themselves where the other costs were concerned.[84] Even so, their apparent decision to withhold even the boarders' tuition fees suggests both a deep hostility to Thring's plans and a dereliction of their responsibility to the school in the face of emergency.[85] Thring tried to see it all in more positive terms, writing in his diary: 'I feel so grateful at the deliverance from the town and the having time once more at our disposal. It is like an escape out of prison. Things may be hard at Borth; there must be much difficulty, but it is the hardness of liberty, not the close deadly grip of a prison'.[86] Even so, there was no disguising the fact that effectively he was to be on his own.

Events now continued to move at great speed. Barely three weeks after Campbell first suggested that they might 'flit', Thring and his staff left the town. They had very little time to pack up any personal possessions and equipment which would not be going with them to Borth, and to arrange for their houses to be looked after while they were away. Thring and Christian had also made arrangements for further tests to be done on their house water supplies.[87] The Lower School pupils remained in Uppingham, despite – or perhaps because of – all the traumas of the previous autumn. We cannot be sure whether Hodgkinson himself decided not to go with Thring, sensing either that it was the wrong decision, or that younger boys were too vulnerable to be uprooted from familiar surroundings, or whether Thring persuaded him to stay in Uppingham because of the shortage of accommodation at Borth.[88] Medically speaking, it seems a strange decision, given the vulnerability of younger boys in particular to this type of disease.

Thus at this point nearly all parties to the dispute had a great deal to lose. Thring and the masters were starting to incur sizeable running costs down in Borth as the school-in-exile took shape.[89] They hoped to be back for the September term with improvements, but this would soon be shown to be a hopelessly optimistic timescale. Even if the reassuring number of pupils who turned up at Borth for the summer term guaranteed the school's future for a while, the masters themselves faced possible personal ruin as they went more deeply into a venture which was unplanned and unbudgeted-for. Thring had started the legal and administrative mechanisms for the new water company, but they were as yet incomplete. He had failed so far to force the RSA to accelerate drainage improvements. He had no means of knowing what pressure the trustees might exert on him in the weeks ahead.

For the trustees the school's absence represented a new financial headache. Whatever their recent pronouncement about being responsible only for the school *at Uppingham*, they were responsible to the Charity Commissioners for its proper administration. The exodus raised the prospect of more work for them, caused by a headmaster who had given them plenty of worry over the years. Their social standing locally would not be improved if the school's fortunes suffered in the long term. It seems highly likely that for many townspeople the prospect of the school's absence brought equally pressing economic concerns. As ratepayers, they had plenty to fear if radical action was taken. A rapid upgrading of Uppingham's sanitation would have a big impact on them, and any piped water supply, whether provided by the RSA itself or by a private company led by Thring or anyone else, had cost implications for the new consumers. Yet if little was done, they faced a crippling loss of trade, as farmers, traders or shopkeepers, through the absence of the town's largest business and employer. Few of them had foreseen events taking this turn, certainly not so soon – and the school's impending departure seems to have made them very hesitant and uncertain. If the school were to be away for only a term, the shopkeepers felt that they could ride out the financial consequences, but a longer absence – perhaps until Christmas – might be a different matter. It is likely that they reckoned that Thring could not hold out that long.

The absence of the school turned the attention of the townspeople powerfully on to the RSA, as it tried to gauge the state of local opinion, and to decide how radically and how quickly to address the sanitary problems. Barnard Smith, Wales and the other guardians thus had to strike a delicate balance, to ensure that ratepayers and traders would not turn on them either for negligent inactivity or for expensive overactivity. They did not dispute that improvements were needed, but they would not be rushed into decisions that they might later regret or which might leave them open to criticism of profligacy if rates were subsequently increased. Details, estimates, tenders and loan arrangements needed to be properly worked out, even if this took time. But they also needed to show that things were moving, albeit slowly; otherwise any combination of the townspeople, the school, the LGB and the press might revive the pressure on them.

Barnard Smith and Wales seem to have had few qualms about exploiting the school's situation, and may even have taken some satisfaction in doing so. They were confident that Thring had over-reached himself. They would wait to see how he fared at Borth. They knew that the trustees were not prepared to put large-scale finance into the scheme, and they would have made rough estimates of the bills that Thring was now running up. Mrs Bell recorded that 'some of the guardians and their supporters were saying that nothing would be done, and Mr Thring would have to bring the school back to the town as he left it'.[90] If he were forced into a humiliating return, either because his funds had run out or because Borth proved to have diseases of its own, the pressure on the RSA would surely diminish, at least for a while. There were some advantages in procrastination, provided that they could keep the ratepayers quiescent in the meantime. The guardians may also have felt daunted, even overwhelmed, by the legal, financial and technical issues which confronted them – and by the fear of making public and costly mistakes.[91] This would explain why they apparently sought to carry out their procedural responsibilities to the letter, regardless of how long this took. They could comfort themselves with the thought that typhoid in the town was no worse than in previous years. They could point to the fact that Hodgkinson and the Lower School had remained in situ, and that his house had not suffered any further illness during the renewed outbreak of typhoid after Christmas. Perhaps things really were improving of their own accord now that increased flushing of sewers and clearing of drains were being carried out.

On the other hand defeat, real or perceived, for the RSA at Thring's hands by rapidly acceding to all his demands might seem to be not merely a humiliation to the guardians, some of whom had given many years of public service: it might also be a threat to their local prestige and influence. Whatever the guardians' view, they were certainly under pressure. Barnard Smith, after so long as chairman, seemed to have wearied of all the months of confrontation some time earlier; on 29 March he had again told the guardians that he wished to retire but 'consented in the present condition of affairs connected with the Union to be re-elected'.[92] The majority of the members of the RSA were happy to sit tight as long as they needed to. Dr Bell and a few others might object, but for the moment at least, they were in a minority. Finally, there were consequences for the LGB officials. For them, Uppingham was just one of many local problems requiring their attention: a comparatively small town with, statistically, a small typhoid outbreak – but one which had already given them a great deal of work. All their instincts were still to avoid taking sides. But there were implications for their credibility and reputation if a desperate school with powerful contacts and a wounded and resentful RSA united to blame them at a later stage. It would prove increasingly impossible for them to remain detached in the months which followed, as appeals and memoranda continued to arrive from both sides.

Dr Bell had remained in Uppingham rather than going with the school to Borth. He could not desert his town patients – and to do so would in all probability have resulted in his practice rapidly being eroded by his two rival doctors. Despite the school's absence, however, he was still its medical officer, and it fell to him to write letters to parents of boys who had previously had mild typhoid at Uppingham and who now needed his permission to send them back to the school at Borth.[93] But he quickly heard of Childs's appointment as school medical officer at Borth,[94] and he realized that his position too might be under threat in the long term. If the school closed, or stayed away permanently, his role would end. Meanwhile it would be Childs, not he, who would have Thring's ear. Bell corresponded with several housemasters at this time to reassure them about the sanitary state of their empty houses.[95] It suited him that Thring had asked him to do this; he was fearful of losing housemaster confidence, and of becoming, along with the RSA, their scapegoat for recent events. He also lobbied Thring regularly by letter in this period for reassurance that his position was safe. His insecurity shows particularly in a letter that he wrote to Thring on 15 April expressing misgivings about the housemasters' view of him, and wondering if some of them might ask one of the other doctors in the town to have this watching brief over the houses.[96] His fears in fact proved unfounded, but the combination of his longstanding loyalty and his insecurity had one positive outcome: it made him highly zealous in the school's cause. He became its main defender in Uppingham, as well as the principal supplier of news to Thring and the housemasters in Borth. His 'Letterbook', along with the LGB papers, is a major source of information. Not surprisingly, it gives prominence to everything that Bell did on the school's behalf. It also reveals an inveterate and caustic letter-writer, as well as someone inclined to see conspiracies at every turn.

The return of typhoid in March 1876 had demonstrated its elusive nature: unlike the previous October, this time no one could reasonably blame inaction by Bell, or point to contagion possibilities between pupils. The key germ and miasma issues seemed to have been addressed, at least within the school's properties, following Tarbotton's report, but residual water problems or a returning carrier could have frustrated them. The town and school authorities were no nearer to the cause, although the disease was almost certainly water-borne. If, in the light of the reports which had now been published, there was any speculation by Thring, the guardians, the MOH or Dr Bell in this period about its origins or the broader epidemiological issues, it has not survived. The RSA had shown great uncertainty, and little urgency, on the sanitation issue in this period, remaining caught more than ever between its desire to make improvements and its unwillingness to face their cost. Its procedural wrangles with the LGB – for example, over details in Rawlinson's report – suggest small-mindedness and a failure to see the true priorities it faced. The LGB itself maintained its low-level-of-intervention stance, although arguably the time for such non-engagement had long passed.

The dilemma facing the RSA over whether to promote or oppose a private water company was typical of utility issues facing guardians in many rural areas: should the guardians see such a venture as a positive step forward or as a dangerous monopoly, open to abuse through escalating water charges? Was there are scope at all for the RSA to run such a venture itself at a profit? Or would such a step merely have increased all the workload and other pressures it faced? Meanwhile the RSA had to face the first ominous signs which had surfaced at the March town meeting of ratepayer concern at its perceived inactivity. Which of its interests would the local 'shopocracy' put first – trading or ratepaying? The former presupposed spending; the latter, continuing economy. And would any promotion of the trading interest be strong enough to overcome that of the landed one, which would still fight to keep rates low?

    The RSA could not assess the impact of the school's absence, or how long it might remain away from the town. It remained to be seen whether its members would rue its lack of initiative and its failure to give comfort to the school over the previous year, whether Thring might eventually be able to dictate terms to them, or if he might be forced to abandon all the buildings he had erected and paid for in Uppingham, allowing them to be repossessed by the trustees for a restored day school – although it is likely that in such circumstances the boarding houses would have become strange and under-used monuments to the past in a town which would have had little immediate use for them. For the school it also remained to be seen whether the move to Borth would be a one which took it out of the frying pan and into the fire,[97] especially if there turned out to be epidemic disease in coastal Wales too. It was conceivable that Thring would eventually be forced into a humiliating retreat back to Uppingham with his sanitary and water supply aims still unachieved. Some members of the RSA believed this to be a possibility, although it is unlikely that Thring ever considered this to be an option. But he must have wondered in his darker moments how likely it was that the school would survive at all.

7 SUMMER 1876

On 26 March Thring sent a telegram from Borth to Christian, who was still in Uppingham: 'It is flat treason and treachery. I have wired to stop it.'[1] We do not know to what this refers, but it shows that distance had done little to dispel his resentful and angry mood. There were also new anonymous letters from each side in the press; 'A father' wrote at length to the *Manchester Critic* at the end of March, complaining at the lack of urgency being shown.[2] This prompted 'One of the townsfolk' to write to the *Stamford Mercury* raising the question of whether Borth would really turn out to be safer than Uppingham.[3] The first edition of the school magazine for the summer term fanned the flames. It included a clever poetical satire: 'How I came to Borth', with the words:

> Leave bickerings and cesspools far behind,
> Take thy stern future with a quiet mind.
> Better are herbs and peace, be well assured,
> Than all the Local Sanitary Board
> Weigh dilute sewage 'gainst pure mountain springs,
> Weigh unflushed drains 'gainst air the salt sea brings
> Weigh all the chances well with equal scales,
> Since Wales won't come to you then go to Wales ...[4]

It did not take long for a copy to find its way to Uppingham, where the rector took offence at the use of his name in this play on words. Dr Bell wrote to Thring on 5 May urging him to stop the boys writing such things; they would not help, especially at a time when he sensed that public opinion might just be starting to move in favour of the school.[5]

All these irritations gave an added spice to the annual elections to the RSA which were due in late April. The elections offered both sides a chance to test local opinion, but also exposed them to potential rejection at the polls. For the school in particular, despatching some old opponents and getting new blood on to the RSA was an attractive prospect. Bell wrote to Jacob that the election would be a close-run thing, but he had identified some possibly vulnerable members,[6] and both he and local solicitor John Pateman would be standing for election. As election day approached feelings ran increasingly high, and so-called 'race pamphlets' were produced, consisting of anonymous reports on the election and its likely results, with nicknames such as 'Blue Pill' for Bell himself and 'Little Awkward' for Barnard Smith.[7] Each side was determined to exploit the voting regulations to the full. Brown, whose role as RSA clerk included running the election, intended to make no allowance for the masters away at Borth, when deciding the length of time which must elapse between sending out voting papers and holding the count. Thus the masters risked being disenfranchised. One housemaster (Candler) had written to the LGB about this before leaving for Borth; he had warned of the logistical difficulties of voting from there, and made it clear that the masters were 'exceedingly interested in the outcome'.[8] The LGB replied that it had no authority to intervene; the clerk, as returning officer, had authority to determine the necessary arrangements.[9]

The RSA decided to send out the ballot papers to the school houses at the last possible moment allowed in law, counting on the fact that it would be all but impossible for the absent masters to cast their votes. Thus the election became a tactical battle of wits. Bell got wind of Brown's plan and was resolved to frustrate it. Supporters of the school followed Brown round to each empty boarding house as he delivered the voting slips, collecting them up and passing them to Charles White, the ironmonger.[10] Joseph Woodcock, arguably the most active guardian opposed to Barnard Smith,[11] provided a dogcart and a pair of horses, and White was taken straight to Rugby station where he caught the last train of the day to Borth. It was a slow one, and he had to travel all night, but on arrival at Borth in the early morning he found Thring and all the masters on the platform with tables, pens and ink at the ready. Mrs Thring had brought breakfast for White to eat on the platform, and within a few minutes (the train having gone down to the terminus at Aberystwyth and come back again) he was on the return journey to Rugby. There he was met again by Woodcock; they managed to hand in the voting papers in Uppingham with fifteen minutes to spare. White's journey proved to have been very worthwhile. After a few days of dispute over a number of doubtful or spoilt ballot papers, several opponents of the school were voted off, and their replacements included Bell.[12] The triumphalist description of White's exploits shows just how sweet a victory this was for the school.

Each side would still go to great lengths to thwart the other. Bell used his new status to become a sharp thorn in the flesh of those guardians who were happy to see the school suffer. He challenged the size of Brown's salary as clerk.[13] He lobbied hard to speed up the formation of the water company and the drainage improvements. These weeks also saw a battle over the plan for a private water supplier, as Bell threatened a legal challenge against the RSA's expenses incurred in opposing the water bill. He demanded that the government auditor surcharge the guardians themselves.[14]

The RSA had been concerned about the continuing progress towards the formation of the water company for some time; its completion would not only highlight the town's failure to provide piped water earlier, but would bring into existence a provider over which the RSA had no technical or legal control. Brown was instructed to do what he could to get clauses written into the Water Bill, protecting the RSA's interests. It was also concerned not to pick up any bill for roadworks caused by pipes being laid. However, it had no positive ideas about how to tackle either issue: Brown was instructed merely to ignore or stall on both matters. Wales told Bell that 'there was no hurry about it, and that the delay was of no importance'.[15] The RSA also asked Rogers Field to come again from London on 4 April. Field reported to the LGB that experimental drilling by the company of new wells to the south of the town was moving forward.[16] He had grave doubts, however, about whether the company would find a sufficient supply in that area.[17] His pessimism proved correct. The company dug to a depth of 420 feet, but it could not obtain an adequate and reliable supply. The RSA seized on Field's doubts to query whether the company's plans would provide sufficient water for hydrants to ensure regular flushing of the sewers and to provide fire protection – thus effectively turning the school's own argument about inadequate water supply back on itself.[18]

Bell wrote to warn Birley of what he saw as mischief-making, but this action unfortunately backfired. As the water company began to consider other possible sites, Birley unfortunately let slip a mention of its activity in Wales's hearing at a trustees' meeting, revealing that further boring work was planned much nearer to the town, in the area between the sanatorium and the town workhouse to the north-west. Wales, whose obligations as a school trustee and as a leading RSA member must by now have been deeply in conflict, put his interests as a guardian first. He passed the information on to Haviland, who stated forcibly that this new site was far too near to sanatorium cesspits which had been so roundly condemned in his report a few months earlier. Haviland conceded that these pits had been emptied and then filled with quicklime, but they had not been dismantled altogether.[19] By this point, however, the RSA had seen the dangers of procrastination and was keen to put up rival proposals in case the water company failed to deliver its promises or to keep its timetable. It asked Field to go beyond merely commenting on the company's plans, and to investigate the feasibility of its producing a rival water scheme based on local springs. Field told it that a great deal depended on rainfall projections; he estimated that between 50,000 and 60,000 gallons per day would be needed. He believed that he had found sources that were pure, but he was less certain of their volume. His experiments were not yet complete; he hoped that steam pumping could be avoided, but if individual steam pumps were needed at several sources, the costs would be too great.[20]

Bell kept abreast of all these developments and acted at times as a go-between for the RSA and the Worksop-based solicitors, Hodding and Beevor, who were acting for the water company. He produced a number of letters with recommendations on the nature of the company share issue, and on the progress being made with trial borings.[21] He wrote to Beevor asking for assurances to be given to the RSA that the company's water price levels would be reasonable.[22] He was also in contact with Mullins at Borth by post, using the latter's keen interest in meteorology to produce rainfall statistics for the weeks earlier in the year. These might be helpful in allaying Field's doubts about the sufficiency of the supply at the site now being proposed.[23] But before Field's researches on behalf of the RSA were complete, the LGB made up its mind decisively in support of the private water company scheme. Forced at last to choose publicly between the school and the RSA, and despite Field's careful appraisal of the technicalities, it concluded in an internal memorandum dated 7 May entitled 'The Reasons for Favouring the Uppingham Water Bill', that the RSA had in fact been opposing the bill without any statutory power to do so. It believed that the Corporation Municipal Funds Act, which had been passed for the purpose of giving such powers to local authorities, applied only to urban authorities, and not to their rural counterparts. This was a rich irony, in view of the RSA's repeated demands over many years for USA status. The LGB also sensed that the RSA had shown enthusiasm for Field's ideas only as a spoiling operation against the private company. It therefore ignored the RSA's legitimate fear that the company's plans might be technically inadequate. The company had (it said) offered to protect the RSA's interest in any reasonable way. There was no guarantee that Field or the RSA could produce a viable alternative scheme, and with the school threatening to remain away from Uppingham until a water supply became a reality, the LGB felt that the bill should go ahead, despite the difficulties which the company was starting to experience on its site at the south end of the town.[24] Rawlinson, true to his habit of favouring local autonomy,[25] dissented from the LGB's view, feeling that it should support the RSA, despite its previous criticisms of him.

The RSA reacted strongly to this decision, sending a deputation to the LGB on 13 May. Bell had been keen to be a member of this group, not because he agreed with the opposition to the bill, but out of mistrust of what Barnard Smith, Wales and Brown intended.[26] In the event he seems not to have been included, although he was able to glean information on its return. He wrote to Thring a day or two later, claiming that the two clergymen had told him that they were confident that they would win the LGB's support, but to his great satisfaction they had been shocked to find a cold reception from the Board, which had told them firmly to put their house in order.[27] The LGB's intervention was decisive in ensuring that the waterworks suffered no further delays. As the time for the bill to proceed through Parliament approached, there was a final, robust debate amongst the RSA's members about how long and how strongly it could be opposed. In the end they decided, either through shortage of funds or expertise, or a belated recognition of the town's interests as trade declined, that they could not produce convincing alternative schemes to Thring's private one. Thus, despite Haviland's opposition, the Uppingham Water Bill had its third reading in the House of Commons on 3 July and received the royal assent on the 13th.[28] It gave Thring and his four fellow directors (including Birley and Jacob, plus John Hawthorn as secretary),[29] the power to raise capital of £6,000 through shares of £10 each, make borrowings and levy charges up to specified limits. The company had a year in which to provide a proper supply at an appropriate site, after which its power would lapse. Work could now begin in earnest.

Sewerage improvements proceeded equally tortuously. It quickly became apparent that these would take far longer than the period of a single school term which Thring had envisaged as he planned his exodus. By early May, Field had lodged his outline sewer proposals with the LGB and Rawlinson had approved them.[30] These included the replacement of manhole covers all along the High Street and the installation of flushing boxes at a number of key points, sections of piping laid at greater depths, repairs to parts of the existing systems and new branch sewers between High Street East (via Queen Street and Adderley Street) down to the large south sewer below the cemetery to the east of the London Road.[31] Field had stated that he would not be able to complete the drawings for at least six weeks, and it was calculated that sewer works could not begin for three or four months. Tenders would have to be invited and scrutinized, sureties produced, loans agreed, contracts drawn up and contractors' plant hired. Bell disputed this time-scale; he also believed that Field's proposals were based on some very high assumptions about rainfall statistics, and feared that these might become a pretext to slow things down still more.[32]

At this point there was a further delay after the LGB announced that it would send a medical inspector to check on progress. This was prompted by concerted pressure from Lord Gainsborough and Sir Charles Adderley[33] who, as the two largest land and property owners in the town, had sought again to use their influence on the school's behalf. They had sent a three-page formal petition to the LGB urging it to investigate the dilatoriness of the RSA, recounting the reasons for the school's move to Borth and stating that it was essential to have sewerage works completed – ideally by the end of the summer holidays, or by Christmas at the latest.[34] The RSA was stung by news of what it inevitably saw as outside interference and another attempt by Thring to exploit his rich and influential contacts. It returned to the attack, demanding the LGB's full support, and claiming that it had been misrepresented. It repeated its protest that its representations about the early publication of Rawlinson's report appeared to have been ignored. It noted with satisfaction Sclater Booth's spirited recent defence in Parliament of the RSA at Eagley[35] and hoped that it could expect similar support. The LGB patiently emphasized its neutrality and called for greater harmony in Uppingham.[36] Sclater Booth, sensing that things were collapsing into recrimination again, decided that this was not the moment to despatch an inspector after all, and wrote to Simon on 17 April: 'I particularly wish that Dr Power should not go down to Uppingham at present'.[37]

Despite this concession the LGB was not going to allow itself to be dictated to by the RSA, which it believed was side-stepping the main issue.[38] It suspected that ratepayer opinion would sooner or later start to shift against the RSA. It had also received demands for action from another Liverpool father, a Captain Withington, on behalf of his fellow parents.[39] Withington's intervention served to highlight not only parental concern that the school should be allowed to return to its roots as quickly as possible, but also a strengthening in parental support for Thring after all the criticism of the previous autumn. Now that the great Borth experiment had been brought to fruition with virtually all the boys safely installed there, the venture was acquiring the status of an imaginative, even heroic, act in the face of small-minded local bureaucrats. The fact that Withington's call, like Rayner's a few weeks earlier,[40] came from Uppingham's recruiting heartlands in north-west England must have given Thring strong encouragement. Withington was about to launch a 'Borth fighting fund' to help Thring with the worrying costs of the school's new location. He demanded to know how he could legally force the LGB to intervene. Faced with this unexpected development, the LGB decided that it would now try harder to get things moving. In a strong reply to the RSA's protests,[41] it declared that, whatever the rights and wrongs of the past, it was now the duty of the RSA to get improvements under way without delay. The Board was dissatisfied with the fact that plans and estimates had not yet been received, and said it was receiving complaints about the RSA which it hoped it would not be necessary to investigate. Unabashed, the RSA retorted that it would 'not venture to express an opinion on the vexatious character of the interference to which they have been subjected throughout the discharge of their duties in very difficult and unexpected circumstances', and that it would welcome a full enquiry into its dispute with Thring. It did, however, send a report from Field on the latest situation, and promised to send yet another deputation to the LGB within a few days.[42]

Meanwhile Sclater Booth added to the pressure on the guardians by his reply to a parliamentary question on 4 May. The *Stamford Mercury* quoted Hansard at length:

> In reply to the Rt Hon Gerard Noel, Mr Sclater Booth said: 'My attention has been called to the unfortunate circumstances which have led to the withdrawal of a well-known school to the coast, and during last winter, [to] the urgent request of the school authorities. I rendered such assistance as I could in the difficult and painful

position to which they were placed, without however presuming to express an opinion on the points of controversy between themselves and the town authorities. I have now every reason to believe that the sanitary authority is ready and willing to undertake such works of sewerage and water supply as are required to put their district into a satisfactory state, and that they have taken the necessary steps with that object' (Hear hear).[43]

It would be hard for Barnard Smith and Wales to ignore such expectation. With Uppingham's affairs once more the focus of its attention, on 13 May the LGB belatedly approved Haviland's reappointment as MOH – a decision for which the RSA had been asking since early February.[44] Bell thought Haviland's reappointment was deplorable, but inevitable; he realized that Uppingham had only one vote in the affairs of the fourteen places in the Northampton combined district which employed the MOH.[45]

Thereafter, a long correspondence began again between the LGB and the RSA over further side-issues: the bye-laws question, the cost of printing the reports of Haviland and Field, how the RSA should pay for it and how the sums should be entered in the accounts. The RSA assured the LGB that in the matter of sewerage improvements it was keen to accelerate the usual tender procedure, and to use 'a local contractor of standing'.[46] The LGB finally received Field's plans and estimates on 3 June, and authorized in principle the loan to pay for them, but it decided on 7 June that a local enquiry should be held before the loan was confirmed. Notices advising ratepayers of the loan should be posted in the usual way.[47] The enquiry would examine not only the case for the improvement loan itself, but also broader questions about the state of the town's sanitation.

Major Tulloch, the inspector, eventually arrived to carry out his enquiry on 7 July. Even allowing for the glee with which Bell described the RSA's discomfiture in a subsequent letter to Thring,[48] this seems to have been a very difficult visit for Barnard Smith and his colleagues, and tempers quickly became frayed. Tulloch first took exception to the fact that the RSA had provided him with some, but not all, of the previous winter's reports on the crisis. It had sent him Haviland's report, together with its own views and the report commissioned from Rogers Field, but had omitted both the more critical LGB report from Rawlinson, and the school-commissioned report from Tarbotton. Nor did Tulloch respond sympathetically to Brown's procedural objections about the advertisement process in respect of his enquiry. When Tulloch went out to see the evidence, it was a hot day. This must have given great encouragement to any convinced miasma theorists: According to Bell: 'The drains luckily stank on that day their best. Major Tulloch said the state of the place was a scandal and that the works must be done. His duties took him to many queer places, but he had never been in one so openly foul'.[49] Thring wrote in his diary that 'Sundry of the townspeople ... spoke pleasantly of the school, and money statistics were advanced without contradiction to show how much the town gained by the school'.[50] Not long afterwards, Bell was expressing concern that the RSA would delay things still more, resentful of the fact that the LGB was insisting on open competition for the tenders, even to the extent of including firms from beyond the immediate locality.[51] The RSA also objected to a proposed bonus for the successful tenderer if he completed the work during a specified period. Once it was known that the school would not be returning in September, the RSA slowed the process right down (apparently in a fit of pique at Major Tulloch's caustic remarks), deciding to re-advertise and then finding that its own form of advertisement was declared invalid on a legal technicality.[52] Thus, on the issue of prevarication Bell's suspicions appear to have been well justified: the RSA would try to place Thring under maximum financial pressure.

Barnard Smith and his colleagues had judged Thring's situation with some accuracy.[53] Once the initial exhilaration of setting things up at Borth and seeing the boys arrive had worn off, Thring's moods became more variable again, and as the long days of June sped by, his diaries show that euphoria was again punctuated by black depressions. In his brighter moments, he was glad to have escaped from the RSA. He explained in a letter to his brother Godfrey: 'I have not had, as at Uppingham for so many years, to sit like Job, scraping boils on a dunghill'.[54] But he could not ignore the pressures mounting on him as a result of the slowness of events in Uppingham and London. A decision would soon have to be made about where the school would be located for the autumn term. He was all too aware that his debts were increasing, despite Withington's fighting fund. Circulars had been sent to every parent, and the fund was publicized nationally in *The Times* on 21 April following a letter to the editor from 'A parent', and locally in the *Aberystwyth Observer* a fortnight later.[55] The *Stamford Mercury* had reported that £200 was raised in the first week.[56] But even this was not enough, and financial worry was never far from Thring's mind. He confided to his diary on 26 May: 'My bank books came this morning – a heavy weight there. I don't quite see how my expenses should be less ....'[57] He raged at his own powerlessness at having had to leave Rutland:

> It has suited the people, who act for Uppingham, to represent us as hostile, but it would be difficult for them to show that we have done anything hostile. As is generally the case when a great wrong is done by people in power, they are lavish of their accusations. My one answer is: 'Why are we at Borth if we are powerful or pugnacious? People are not turned out of house and home and brought face to face with ruin for their own amusement'.[58]

He also knew that he had to do what he could to work with the trustees, even though he had clashed with them so much over the years, and had developed a profoundly pessimistic view of the level of understanding that they had for the school and its achievements. This gloom was compounded by two extraordinary facts: not only had there been minimal contact from them since the school left Uppingham in March, but there had been no visit from any trustee to Borth during this summer term – apart from Birley and Jacob who had been down to see their sons. Nor would any of them come throughout the school's entire stay in Wales. Any knowledge of its situation would be acquired only at second-hand. Yet they were still his employers, and Thring grew increasingly anxious as news arrived that they were to hold a special meeting on 17 June. He wrote to Birley:

> Bear in mind that a fiat of the trustees on Saturday for return, without an affirmation of safety, means the break-up of the present school. If they order, without giving assurance of safety in their judgement, the order will not be obeyed. And I think I may add a large number of masters will stand by me in this refusal ... It is strange sitting here and waiting quietly for one's doom, and at such hands.[59]

All through June he had been testing the mood of the masters about the possibility of the school spending a second term away. His diary entry on 22 June shows that the idea of remaining in Borth had initially met a great deal of opposition:

> The conduct of the masters up to a fortnight ago, [even] in fact up to the meeting [i.e. of the trustees on 17 June] had almost been such as to make me tell Birley and Jacob that to hold to [i.e. stay on in] Borth with such disaffection was impossible ...[60]

He was haunted by the idea that he might be forced into a humiliating return to Uppingham, with proper drainage and water supply still unprovided. In such an event, they would have lost 'as I told them weeks ago, almost all the advantage that we had gained by our daring move and its trials, and that all that could be done now was to avoid unconditional surrender'. Capitulation was unthinkable as far as Thring was concerned. He decided that 'Things tend more and more to a final breaking away from Uppingham'.[61] Undaunted by the sceptics, he talked of refounding the school elsewhere. He was not entirely alone in this; at least one housemaster (Bagshawe) was voicing the same thoughts.[62]

It is impossible to say whether news of any of these doubts and disputes got back to Barnard Smith and Wales, and if so, whether they passed them on to the trustees. But the trustees appear to have been aware of them; at the meeting on 17 June they confirmed Thring's worst suspicions about them by declining to take medical advice about the latest state of the town before ordering him to bring the school back to Uppingham in September.[63] If only the masters had presented a united front for staying in Borth ahead of the meeting, Thring might have been able to persuade the trustees at least to defer a decision. But they had not been united, and now if the school did not return as instructed, he would be in direct confrontation with his employers. He feared they would then have the pretext to dismiss him.[64] In the event, the trustees' stance spectacularly backfired, because, faced with such high-handedness from such a group of men so remote from their situation, the housemasters' mood suddenly started to change. Belatedly they rallied behind Thring. Even those who had been demanding that everyone return to Uppingham as soon as the school broke up for the summer now began to tell him that they thought at least one more term away was inevitable. The diary records that 'S' (probably Skrine) 'is now convinced that we ought to stay here next term, and shall probably have to do so'. Thring added: 'I said this should have been the masters' opinion six weeks ago, when it would have made all things easy; that now it was impossible to move [i.e. to change plan and extend the stay] ...'[65]

By 1 July nearly all masters' were convinced, not least because Dr Bell had written to Thring with news of fresh typhoid cases in the town:[66]

> As I knew you had to give a decided answer to your landlord [at the Hotel] at the end of this month, I thought it best to drop a line as to my suspicions that you might avoid giving such answer as you might wish afterwards to withdraw ... I fear it must decide you to stop away for the next term, I cannot see how you can come back in the face of it.[67]

As a result, direct confrontation with the trustees was unnecessary. Their June demand for the school to return in September, closely followed by these new typhoid cases, was quickly overtaken by Tulloch's devastating report, together with a strong subsequent message backed by the LGB stating that on no account should the school return before Christmas. When the trustees met on 14 July they had no alternative but to reverse their earlier decision.[68] A telegram to this effect arrived in Borth at 5 p.m. It was expressed in face-saving terms, as the minute book records:

> In the opinion of the trustees there is nothing in the present condition of the town of Uppingham to cause them to rescind their resolution of the 17th ult., yet having regard to a memorial addressed to them by the whole body of the assistant masters they are willing in compliance with the same that the school should remain in Borth during the autumn term.[69]

Thring drew wry amusement from it: 'It is fun to see what a sour face they make over it, and are foolish enough to show that they make'.[70] At least they granted a further £500, in advance of the following term's fees, to keep the masters financially afloat.

A few days later term ended on 'a glorious day, bright and hot'. The boys departed by train – but not before Thring had told them 'to come back with the soldier spirit to face whatever remained'.[71] But Thring was a realist as well as an idealist, and he surely knew that a second or even third term would have none of the novelty of the first. Summer, with warm weather and so many possibilities out of doors, had been a pleasurable experience; a winter term – with short days and variable, cold weather – would pose far greater problems. Birley had been in Borth on 7 June when he wrote a letter to Dr Bell: 'the place is glorious now, but *I do not think it tenable in winter in its present condition* – you need not tell the rector'.[72]

Meanwhile, there was one unexpected result of the school's presence in Wales. If typhoid could be infectious, so could an enthusiasm for public health reform. The people of Borth had begun to focus their attention on Borth's own lack of a public water supply. While the school was away over the summer, a public meeting took place. There was much talk of smells and dangers – and much raking over of old complaints about how the Cambrian Railway Company had worsened the situation by altering the course of local streams near the village when building its embankments a decade earlier. If the local RSA could not, or would not, provide it, other means must be found to finance a waterworks.[73] Similar meetings were held in Aberystwyth itself a month later.[74] The *Cambrian News* mused: 'How watering places can expect to flourish as long as visitors are unable to obtain even scanty supplies of doubtful water is a mystery'.[75] Thring, on holiday in the Lake District, was not there to witness such protests, but it is likely when he heard about them that he recognized their familiar ring – and that the irony was not lost on him.[76]

With the summer term at Borth now ended, the masters went their separate ways for the summer break. Thring, in need of rest, departed to his usual retreat at Ben Place on Grasmere in the Lake District.[77] It was left to Christian, the only one of the housemasters who appears to have spent much of the summer in Uppingham, to handle matters on Thring's behalf. Dr Bell would help him to keep things moving as best they could, with Birley and Jacob guiding the effort and giving legal advice where it was needed.[78] Bell continued to relieve his frustration through literary activity – partly by reviving his long dispute with Haviland. When he reported to the RSA that typhoid had broken out again on 1 July,[79] it immediately informed the LGB. Haviland was sent to investigate, and claimed:

> I proceeded to the premises where I met Mr Bell and requested him to accompany me... He however refused to do so and dared me to enter the premises ... Having been thus impeded in the execution of my duty, I left ... and report the fact, asking how I am to act under the circumstances.[80]

Bell immediately wrote to the LGB himself, seeking a ruling that Haviland had no power to enter a private house in such circumstances without the agreement of its occupier.[81] The RSA, in a difficult position now that Bell was one of its members, also decided to ask the LGB for a ruling on the issue. It replied, confirming that Haviland had no such power.[82] Three days later, Bell returned to the attack, pointing out to the LGB that under the 1875 Public Health Act MOHs were required to look into causes of disease outbreaks *as a whole*, but not into individual cases. He claimed furthermore that he had kept Haviland fully informed about this case, despite the fact that Haviland had given no apology for earlier incidents between them. A characteristically cautious LGB memo in reply suggested that it could, or should, not interfere; this was deemed to be a matter of professional etiquette, not of law, and Bell had no grounds for complaint in law.[83] Bell again wrote, thanking the LGB for its support, and justifying himself again at length. It was insulting for Haviland to talk about 'a *supposed* case of typhoid fever'. This infuriated the LGB, which recorded an indignant memorandum against Bell's insensitivity in pursuing the issue so remorselessly despite being a guardian himself. It considered whether 'to advise Mr Bell of his social responsibilities', but decided eventually that 'the safe course is merely to acknowledge it'.[84] Unabashed, Bell then researched further, and wrote to Jacob claiming that he had discovered that Haviland had failed to send in annual reports and illness and mortality returns for either 1874 or 1875: 'If the LGB stand [for] their official leaving their letters unanswered, they will stand [for] anything'.[85] Bell now broadened the issue again, questioning the RSA's every decision. What were its motives in allowing further delay? Was there not a risk that with the project so delayed and so contentious, only small contractors would tender for the work, and that it might be inadequately done? Why was it so resentful of the fact that the LGB was insisting on open competition?[86] Why was it opposed to a bonus being paid to any contractor completing the work during a specified period? He wrote to Thring suggesting that it was surely unreasonable for Haviland to continue to object to the proposed flushing arrangements of the sewer system all through July and August, ostensibly on the grounds that the water company was not yet in a position to guarantee enough water to make them work.[87] Thring, on holiday, appears not to have responded to this.

The RSA meanwhile became involved in a new dispute with the LGB over the terms of the proposed loan to pay for the new sewers. It seems likely that Treasury demands for interest rates higher than 3.5 per cent to be the norm were now being brought to bear on the LGB's officials.[88] Major Tulloch's initial recommendation after the Board's enquiry in Uppingham in early July was in line with the new Treasury policy,[89] but the RSA pleaded that its situation merited being judged a special case, and that its rate should be set at only 3.5 per cent. The LGB agreed to recommend this, but it warned the RSA that it had no power to overrule the PWLB, which might veto the LGB's recommendation.[90] Furthermore, it insisted on putting forward the proposal to the PWLB that the loan be repaid over the newly-required loan period of only thirty years, rather than the fifty years which the RSA wished for.[91] The ratepayers would have to foot the increased bill.

As the day for the opening of the tenders drew near, Bell became increasingly anxious. He wrote again to the LGB: the weeks were slipping by, the summer would soon end and the weather would deteriorate and construction work would become more difficult and expensive.[92] He wished that the LGB would again send down someone from London to force the pace. He believed that the LGB lacked the will rather than the legal power to interfere; he felt that the RSA had watched the recent successful resistance elsewhere of another board of guardians (at Keighley, Yorkshire) to LGB pressure in a vaccination dispute, and were taking their cue from them.[93] He also suggested that Jacob visit the PWLB Office in an effort to accelerate a firm decision about the loan. He hoped that Rogers Field would be present when the tenders were opened to contribute his engineering expertise, although he was sure that Barnard Smith would prefer him not to be. Bell particularly feared that if the water company could not guarantee sufficient supplies immediately to flush the new sewers, Barnard Smith would use this as a pretext to delay sewerage works until the waterworks was complete, possibly by up to a year: 'I said to him, you cannot put off the works until that time. Oh yes, he said, we can, if Mr Field and Mr Haviland tell us they ought not to be done. Barnard Smith does not want to open the tender in that case.'[94]

Bell had become convinced that the RSA was deliberately doing the least it could get away with, and that it was determined 'to make Mr Thring submit to them'.[95] He confided similar fears in a letter to Sir Henry Thring, wondering whether the ratepayers ought to be goaded into action against the RSA – or even whether the LGB should seek special parliamentary powers to override it. It needed to act with 'energy and firmness'. He reiterated his regret at Haviland's recent reappointment as MOH, and wondered if it boded ill for the future.[96] Thring shared Bell's pessimism from afar. He wrote to Christian from Ben Place on 9 August:

> How I hate the whole subject ... The rector has written a specious letter to Mr Jacob which is most instructive. He lets out that since Sir C Adderley and myself have failed to bring them to book with the LGB, no other power can. This is the secret of their insolent security. He is instructively [i.e. instinctively?] blind to the fact that there may be other reasons behind a want of power for not pursuing a matter to the bitter end, and other penalties besides the law.[97]

He hoped Adderley would fight on, but felt that it was not for him and the masters to fight the RSA 'over the [interest] rate'. If the town was not prepared to admit its errors, there was little he could do. He believed that the LGB was threatening not to confirm Haviland's reappointment, although he did not expect the threat to be carried out.[98] On 13 August Bell wrote to Jacob that the RSA was now putting out false information about the increased rates which would result from the sanitary works, partly to cover its tracks for the costs of its opposition to improvements earlier.[99] He saw every delay and every problem as a conspiracy rather than the product of accident or incompetence – even the slowness of a final decision from London over the loan: 'I do not think that the LGB Inspector [Mr Beaumont] was here accidentally, and I think that Barnard Smith knows more about that than he cares to tell'.[100]

Bell had kept up the pressure almost single-handed at a critical time, but August saw a period of sustained pressure on the school's behalf by others too. William Earle, the longest-standing member of the staff by some years,[101] wrote three letters on 14 August. First, as 'the Second Master in Uppingham School' he asked the LGB to compel the RSA to complete the sewerage work by November. He claimed to be writing at the behest of leading ratepayers in the town, a community with which he had very longstanding links, as well as on behalf of the school. Again the LGB stood back, referring the request to the RSA, which responded through its clerk on 28 August that the new bye-laws had been fully published, and were now agreed and adopted.[102] Progress on Earle's concerns could now be expected very soon; it would also shortly be accepting a tender for the sewerage work, subject to the references being satisfactory. The LGB replied, emphasizing that there must be as little delay as possible.[103] Earle also wrote a letter to rector Wales. Their friendship went back nearly two decades, and he hoped it would survive these controversies:

> I can hardly tell you how distressed I am that the Board are again going to postpone the drainage; I simply cannot believe that they will do such a thing. I hope that you will not only not sanction [the improvements], but that you will let your disapproval [of procrastination] be publicly known. It will endanger the peace of Uppingham in our time. I beg and entreat you as one who has been and who still desires to be your friend, desiring as I do the return of mutual goodwill, rendered more uncertain than ever, to do all you can ...[104]

It was a dignified and moving plea, from a cautious and moderate man. No reply has survived.

Finally Earle wrote to Lord Gainsborough, Sir Charles Adderley and Sir John Fludyer, as powerful local influences and trustees, saying that 'No time should be lost'. He sent copies of the letters to Christian. He suggested that a large number of ratepayers should be assembled to go up to London; he would gladly come himself or 'meet the expenses of others to go'. Christian (by then on holiday in Ilminster, Somerset) replied that the news of further delays was indeed 'disgusting and really alarming ... I am more and more of the opinion that the time has come for a more distinctly aggressive policy on our part'. He approved of the idea of a petition to London and a deputation if necessary. Meanwhile there should be a ratepayers' meeting. He recalled that all the magistrates could sit on the

RSA ex officio and suggested that they too should be contacted to apply their influence.[105]

Meanwhile Christian had himself been active in three ways. On 11 August he too had written to Sir Charles Adderley, encouraging him to put down more parliamentary questions. Adderley replied: 'It is inconceivable that *men* [*sic*] should act thus'. He also sent Christian a telegram: 'Find names of any members of parliament having sons at Uppingham who would ask question in the House of Commons'.[106] Christian appears to have acted on this advice, for three days later, in response to a question in Parliament from Mr Whitwell, a backbench MP, Sclater Booth denied that the postponement of the school's return was due to the non-completion of drainage works.[107] On what basis he made this statement is far from clear. He surely cannot have believed this, but he may have seen it as counterproductive for the RSA to be publicly put under any further pressure. Christian also wrote to Charles Clode, the legal under-secretary of the PWLB, on 16 August, requesting a speedy decision on the loan question, and pleading the case for an interest rate of only 3.5 per cent.[108] Clode replied a day later that he thought there would be no problem with this. Christian also received letters from both Birley and Jacob saying that the LGB should be contacted immediately if the next meeting produced further delays, together with an expression of support from Hodgkinson[109] and a letter from Thring himself with the postscript: 'I am so sorry you have all this worry'.[110]

By now, although there was no obvious end in sight, Thring held the view that the school had done all it could and that it was up to the ratepayers to assert themselves. He believed that they would do so, because the ineptitude of the RSA was self-evident: 'The utter want of business acuteness makes one laugh ... clever men would not bungle so much in conducting their own case'.[111] He broke his holiday and went down meet Birley (and Jacob) in Manchester; afterwards he wrote back to Christian: 'I quite agree that the crisis seems to have come, but I cannot think that the school in my person should be dragged through the mire of a street fight with the rector ...'.[112] Birley observed:

> I am sure that if the inhabitants of Uppingham care for the school to be amongst themselves they must assert themselves as they have never done yet – I find parents of boys here [in Manchester] very little inclined to lend any help – they argue that if Uppingham does not care for the school they need not have it – and that it would be much better if Mr Thring would leave the place and set up his flag elsewhere.[113]

He also suggested that the time might have come for the LGB to undertake the sanitary works itself.[114] His view that maximum pressure should be exerted on the RSA was backed up by Mullins, also on holiday in Somerset, in a letter to Christian on 14 August:

> The intelligence you give is disgusting and really alarming. I think you have done
> very well in getting so decided a resolution passed, and I shall be glad if you will add
> my name to it ... I do not think any time should be lost in having the petition to the
> LGB prepared ... I am more and more of the opinion that the time has come for a
> more distinctly aggressive policy on our part. Everyone I meet or hear from speaks
> in the same terms. It would be a great thing to get someone of independent position
> to attend the sanitary meeting on Wednesday. I believe all magistrates residing in the
> town are ex-officio members.

He suggested enlisting the support of two other local clergymen: Revd John
Piercy of Slawston or Revd Harry Upcher at Allexton.[115] He thought that Bell
knew Piercy:

> The men who are afraid of the LGB might rally to a leader who was not afraid of
> B[arnard] S[mith] or the r[ector] ... I will willingly find £5 (or if necessary £10)
> towards the expense of retaining good legal advice at this crisis. I think at any rate
> that Haviland's power to interfere should be questioned.[116]

The 'crisis' to which Thring had referred was a renewed demand by local residents
– at last – that Barnard Smith meet a ratepayers' deputation. The local tradesmen, having served partial notice of their frustration four months earlier,[117] were
now asserting themselves in earnest. A deputation of ratepayers had been assembled, representing no fewer than seventy-five others.[118] The memorial they had
drawn up pulled no punches:

> We the undersigned ratepayers believe ... that our interests will be seriously damaged
> by any further delay in improving the sewerage: that any addition to the long delay
> that has already elapsed must add heavily to the pecuniary loss, inconvenience, and
> suffering which many of them have already undergone, and will imperil the existence
> of the school upon its present important scale, and prove a deep and lasting injury to
> the ratepayers and owners of property in the parish.[119]

The deputation was led by John Hawthorn, who, as a printer and bookshop
owner, would have felt the school's continuing absence as keenly as anyone, and
who would have had much to lose personally if it left Uppingham permanently.
His principal supporters were William Compton and William Garner Hart.
Compton's intervention was highly significant; it was he who had led the call for
improvements as early as 1857,[120] and as churchwarden and a prominent church
benefactor, he was one of the few town traders who was not a dissenter. He was
perhaps the only person who could have called Wales to order. That he did so
publicly, suggests that his patience was at an end.[121]

The deputation's meeting with Barnard Smith and Wales took place on 13
August, in circumstances which Bell later described in a series of letters to Jacob
with some satisfaction: 'The deputation attended at 11am to present the memorial [demand], and it was then arranged to meet them again at 3pm to state what

had been done [recently]'. When Barnard Smith stated that the RSA was dropping its objections to the water company:

> The deputation expressed themselves 'perfectly satisfied' [but] then the rector [Wales] allowed his temper to get the better of his judgement, and said attacking Mr Hawthorn, 'that they were not to suppose the memorial had made the least difference to their decision, and no memorial could exert any influence'. He was going on in this strain when Compton said: 'Come Mr Wales, don't spoil it, we are all harmonious now', others joined in so the rector shut up, contenting himself with telling Mr Hawthorn, 'that he hoped now he would use his best influence to bring about a more charitable and peaceful feeling in the parish'. Hawthorn replied 'he should leave that, to someone more influential than himself', the memorial was too fully signed to please the rector and his friends'.[122]

Bell added that there was to be another meeting if necessary a week later; he thought the deputation would decide to contact the LGB immediately after that, if there was no further progress. He also passed on to Jacob the fear voiced by Christian that the rector and others might make trouble for some of the leading signatories of the deputation – although precisely how he thought that this might be done is not recorded.[123] Christian's anxiety was probably unfounded as things turned out; with the RSA in disarray, Field appears to have joined in the backtracking from questioning the ability of the water company to provide enough water for sewer flushing. This change of heart had greatly lessened the impact of the opposition which Haviland had again expressed about the company's new drilling site near the sanatorium.[124]

Three days later the tenders for the sewerage improvements were opened. Seven had been received, ranging between £1,800 and £3,300. The lowest was from Mr J. H. Smart of Northampton. Field agreed to examine them, pending a further meeting in a few days' time. 'No men with the slightest particle of integrity can get away from that, still there is a loophole I should like to have stopped ...' wrote Bell. He was keen to hear definitively from the PWLB that it would agree to a 3.5 per cent interest rate.[125] A £3,000 loan was confirmed at 3.5 per cent over thirty years; it would prove to be more than enough to do the work, but its size, even at the lower-than-usual interest rate, would weigh heavily on the ratepayers.[126] Anticipating that Smart's tender, the lowest, would be successful, Bell also wrote to a Mr Cogan in Northampton to seek assurances about Smart's suitability; he did not trust the RSA's clerk to do so with any urgency, and it might prevent further delays if he could himself ensure that a recommendation was ready for the next meeting.[127] He reported back to the RSA that all seemed well, and it passed the news on to the LGB, which replied that there should now be a rapid start to the work. Smart's tender for £1,864 was accepted on 23 August,[128] four days after another angry meeting at which ratepayers, again led by John Hawthorn, protested against all the delays – and Wales again lost his temper.

Thring had little sympathy for Barnard Smith, although he saw Wales as the villain of the piece. He wrote to Christian again from Ben Place on 17 August: 'The rector is just like a naughty little boy crying "I don't care, I don't care!" when put in a corner. I am sick of his cant about "controversy" and "our not joining them".'[129] A week later, he wrote again. He was gaining 'rest and strength' in the Lakes, fortified by the belief that:

> We have now entered on the last scene of the curious drama that has been playing this year, and I trust it will be played out well ... Nothing surprises me in the rector; he has clearly got out of his depth, and his nose full of water, and may splash about a good deal. I hope the masters at Borth will treat him with cool civility if they see him.[130]

But despite the obstructiveness of Barnard Smith and Wales, and continuing concerns over the financial and technical issues, some members of the RSA were keen to press forward and to make peace with the school. It is possible that a preoccupation with the impending harvest was turning their attention to other pressing matters: we cannot be sure. But whatever the reason, Christian received a revealing letter from a leading guardian and farmer, Edward Wortley of Ridlington, on 17 August. Wortley professed not to have been fully aware of recent events or the latest stormy meeting – something which seems surprising in view of his long experience as a guardian. He pointed out that some of the delays over the sewerage question had been 'Partly legal and unavoidable hitherto', but that he believed 'now to hold back or defer or not to urge on with all speed would be childish and cruel'.[131]

It was a welcome gesture – almost the only olive branch between the two sides in the entire five-month period since the school had departed. Things could now surely only get better, but it remained to be seen just how long it would be before the school could safely return. Earle wrote to Christian on 19 August: 'All will I trust now go smoothly and oh! For the return of peace and happy days'.[132]

In the summer months, the RSA had been at its most deficient in terms of providing leadership for the town, even in the face of the school's departure. It had taken up a defensive and fearful posture towards both the local elections and the water supply question – in the latter case harassing the private initiative but failing successfully to promote one of its own, until eventually there was belated but decisive intervention from the LGB. The RSA had procrastinated over sanitary improvement despite evidence of the need for urgency which Major Tulloch's visit confirmed – out of a combination of the enormity of the problems facing it, its inability to prioritize or to tackle them, fear of making technical, legal or procedural mistakes, confusion and wariness about the likely local trader reaction to the choice between continuing caution and lower rates or radical reform and higher charges, and spite against Thring himself. It remained to be seen whether its fears about the new water supply would prove correct. It had taken the combined pressure of the petulant but effective Dr Bell, the ratepayers and the LGB to achieve results. Recognizing that the absent housemasters could not this time play the leading role that they had taken at the March meeting with Barnard Smith and Wales,[133] the tradesmen had now themselves confronted the RSA – with the LGB following suit. Local pressure continued to be the main driver of its decisions – and local trading conditions would drive sanitary reform, rather than vice versa.

Amid all this controversy, one fundamental mystery remained: what had caused the typhoid outbreaks which had given rise to all this conflict? With the school away at Borth, the disease was now confined to a small number of cases scattered across the town. But it was still present, making it at least possible that a carrier who worked in the school but who lived in the town had been the original source of the epidemic. There had been sporadic town cases for many years. But on this issue there could be only speculation – and it seems that the leading figures were enmeshed in too many other controversies to spend much time indulging in it.

The masters had been won round to the necessity of a second term in Borth, and even the trustees could see that the school could not yet return. Thring had won this latest round, but his financial and other worries remained very real. Moreover the animosity between school and RSA, headmaster and trustees, MOH and school doctor, even RSA and the LGB, remained deep – and any attempts to build bridges between the warring factions were still very fragile.

8 AUTUMN, WINTER AND SPRING 1876–7

The wishes of both Wortley and Earle for more harmonious times were not yet to be realized. The fragile truce after the ratepayers' revolt was soon tested as the controversy resurfaced in the national press. The anonymous 'Paterfamilias' returned to the attack with an inflammatory diatribe in *The Times* on 28 August. Reminding readers of all the events and specialist reports of the past year, he stated that the school had 'at once carried out all [Rawlinson's suggestions] with unsparing care'. By contrast, the plans of Rogers Field for the town improvements 'were adopted by the local sanitary board ... but no effectual effort has been made to carry them out'. Warning that there was still no guarantee that the school would be able to return even after Christmas 'unless more activity is displayed in remedying the original evil', he called for an end to 'mischievous and harmful delay', criticized the trustees for being supine and described:

> ... the spectacle of a great school under a man of admitted originality and power ... whose masters had spent over £80,000 on buildings over the previous twenty years ... driven from their rightful home to an obscure Welsh village, at the extremity of the land, leaving, like the old Phocaeans 'their fields and Penates and beautiful Temple' to lie waste and desolate.[1]

Bell wrote to Jacob that this had:

> acted like a blister, and some of the Authority were very unhappy about the 'lies' it contained. It was debated whether there should be a reply, but it was thought best to leave it alone, because 'while the school can get fair space allowed in the *Times* for anything they have to say, the S.A. would have their letter mutilated and pushed into a corner'.[2]

But 'An old inhabitant' did decide to reply. He wrote to the *Stamford Mercury* on 1 September, listing the small number of deaths in the town in recent months. This, he believed, showed that it really was a healthy place,[3] and that the RSA had done all it could. He pointed out that 'the Uppingham School was founded for the benefit of town and district' – from which he concluded that:

> Paterfamilias and other parents with large families, to whom a good and cheap education is of great consequence, take advantage of our charity and send their sons to

reap the benefits, and are the first to raise an unjust cry against the town from whose charities they have received and are receiving great benefit. I would ask you to allow me to draw the attention of the public to the fact that a sum of £2,000 was expended four years ago for drainage; that a rate of 2/8d in the pound has been laid to pay for expenses in investigating the causes of the outbreak of fever ... and that the rates are close upon 10 [shillings] in the pound, which will all be largely increased when the drain is finished, and this falls solely on the owners and occupiers in the parish, and Paterfamilias pays nothing towards the expenses that he so loudly calls for.

He also criticized the school for having failed thus far to provide a water supply 'from want of capital, energy or proper advice'.[4]

Meanwhile, it had been one thing to move the school to the Atlantic coast for three months at the start of summer, but quite another to return there as winter set in. While arguments continued in Uppingham, the masters in exile had one immediate concern: the need – literally – to batten down the hatches in Borth.[5] The cost of making the school's makeshift accommodation suitable for continuing occupation at least through to Christmas was of great concern to them. It was only partially offset by another payment (£169) from Captain Withington's fighting fund.[6] The fund had now contributed nearly £730 – which the *USM* later estimated, surely with extreme over-optimism, to amount to about a third of the whole cost of the project.[7] There was a further worry too: the discovery of seven cases of scarlet fever amongst the boys at Borth on 26 October. However, Childs imposed stringent isolation and the outbreak was over in ten days. The *Lancet* (25 November and 2 December 1876) stated rather sourly that Thring would not be able to blame the RSA for this outbreak, and compared Uppingham's disease record and care arrangements very unfavourably with those of Marlborough.[8]

In the early weeks of the new school term it was assumed that the pupils would be back in Rutland soon after Christmas.[9] This expectation resulted from news which arrived in Borth on the first day of the new term on 15 September that sewerage work in Uppingham had begun at last. It was also reassuring to Thring and his staff that Hodgkinson had experienced no further problems in the Lower School in the months since Tarbotton's improvements had been carried out there.[10] As the weeks went by, however, typhoid again reappeared in the town,[11] the works proceeded disappointingly slowly and the date of the school's return was again put back. While masters and boys struggled through a Welsh coastal winter, controversy remained undiminished within Uppingham. Only with the arrival of spring and the completion of the sanitation and water improvements could it safely return.

After all his campaigning in the summer, the autumn brought only limited respite for Dr Bell. He believed 'An old inhabitant' to be a former member of the RSA – possibly one of those voted off it earlier in the year: 'It is a great pity that they do not stick to the truth. They are like the ostrich, they cannot see their deficiencies and believe everyone else is blind.'[12] He wrote to the LGB: 'I have told this Authority that I am anxious to assist them in carrying out their duties. But that when they endeavour to exceed them, I shall oppose them'.[13] The RSA had indeed set a new, higher rate, which Bell still believed could have been avoided, but for all the disputes which it had precipitated in recent months.[14] He was concerned that it might now aim for still more delays, in an attempt to phase the rising costs of the sanitary work. Haviland's description of Bell's conduct over the June typhoid cases still rankled; there was another typhoid case in the town on 19 September, but this time he would not risk further trouble by reporting it. He justified himself by launching into another denunciation of Haviland: 'One asks: What is the use of a medical officer?'[15] He wrote at length to Jacob about the sewer progress – and about Haviland.[16]

The *Stamford Mercury* reported on 22 September that Mr Smart had begun work on laying the drains. Even now things did not go completely according to plan:

> On Monday evening, as Mr Holman of Bisbrooke was returning from Leicester, one of the holes being left unprotected, the horse got in and injured itself severely, breaking the harness. Fortunately the occupants of the cart escaped unhurt. On Tuesday evening, Mr Askew went to look at the place where the horse slipped in, and by some means he got in and sustained serious injury.[17]

Bell kept Thring posted about sewerage developments. In a letter of 6 October he detailed progress in great detail: digging deep was proving harder than expected and was also likely to lead to the work taking longer than anticipated. He spoke of 'miscalculation and blunder' as well as four more typhoid cases among his own patients and rumours of several others elsewhere in the town.[18] This was news which Thring must have received with alarm as the autumn evenings drew in. He recorded in his diary:

> We hear that the drain work has brought some fearful revelations, and that the chairman has had to come and see to it, as the workmen refused, near the workhouse, to keep on the whole day. I grieve to say there is more typhoid going on there. I suppose at last their eyes must be getting opened up, but I don't know … The popular feeling at Uppingham, if not stirred up, must gradually find out that we have been most patient, instead of aggressive.[19]

His foreboding proved well-founded. On 29 November, Smart, the drainage contractor, applied to the RSA for extra time to complete the sewer work. Bell did not think it was Smart's fault: rather that the problem lay in poor projections in the tendering.[20]

Meanwhile Bell had a new issue to pursue with Barnard Smith and particularly with Wales. He wrote to Christian on 15 November that he had drawn the attention of the RSA to the longstanding drainage problems at the national (town) school – something which Christian had suggested that Bell might investigate. This triggered a rare disagreement between Barnard Smith and Wales:

> The chairman [Barnard Smith] and two or three others appeared glad to have had the matter brought before them ... they have been at the rector about it before, and he has always asked for time, pleaded that they had no funds, that the Authority ought to help and that the gradients were unsuitable etc etc all to delay ...[21]
>
> Virtually Mr Wales does as he likes in the management of [the national school's] affairs.[22]

Bell did not let the matter go, forcing the board of school managers, of which Wales was chairman and whose number included one or two RSA members, to get estimates for improvements, and demanding wholesale resignations if nothing was done.

Following another confrontation at the RSA meeting on 1 December, at which Wales again pleaded a shortage of funds and claimed that the guardians had no power to force action from the school managers on this issue, Bell threatened to form an alternative board to overthrow or even buy out the existing school managers.[23] He wrote to Thring to suggest that the masters should either subscribe to this campaign (a suggestion probably not received over-enthusiastically, given all the other financial pressures on them) or use John Hawthorn's recent resignation as a manager as an opportunity to get a master elected.[24] He suggested that Wales was actually in favour of this, wishing that his old friend Earle be nominated. Earle had replied, putting down conditions (the nature of which are not known) on any such nomination, which Wales was urged by other managers to reject.[25] Bell felt that 'extreme intimacy with the rector' might place Earle in a very difficult position.[26] Yet he did not wish to take on the role of school manager himself, and he was sure from all his previous dealings with another of Wales's suggested candidates, W. H. Brown, that Brown would be far from ideal.[27] He suggested that Hodgkinson and Campbell would not wish to stand for the role, but thought that Mullins, Candler or Christian might. In the event, the masters decided to support Earle's candidature and he was duly elected.[28] The school sanitation issue then rumbled on for some months; it was established that there was a dry-earth closet system in place, and that these closets would now be treated daily. Earle undertook to keep the issue under review, but some managers felt that dry-earth arrangements would not be a good long-term solution.[29]

Encouraged by the discovery of new areas in which the RSA had failed to discharge its responsibilities, another housemaster, Candler, wrote to suggest that Bell ask some questions about the state of drainage at the union workhouse.[30] Shrewd as ever and perhaps still elated by his foresight in thwarting the RSA over the previous spring's elections, he had thought of something which everyone else who supported the school's case seemed to have overlooked. The workhouse, like the sanatorium whose cesspits Haviland had criticized so strongly a year earlier, was very near the new water supply station. Haviland had been vehement in June about the risks of allowing the water company to choose a location near the sanatorium as an alternative to its first-choice site to the south of the town which had proved so disappointing. The issue of this siting had now been festering for a full six months. Candler now reckoned that the company's supply would be roughly the same distance from the workhouse as from the sanatorium, and he was curious to know about the workhouse drainage arrangements. It might be possible seriously to embarrass the RSA if it could be accused of criticizing the sanatorium pits, while ignoring – or even keeping secret – the state of pits at the workhouse only a few hundred yards away. Bell seized on the issue with alacrity, but his first challenge merely produced evasions. Barnard Smith 'could not say' what state the workhouse pits were in, or whether they were all to be connected to the new system. Barnard Smith then played for time, saying he was due to meet the workhouse master a week later, and would pursue the question then.[31]

Barnard Smith may well have hoped that the recurrence of typhoid cases in the town early in December would divert Bell's attention away from this potentially difficult issue, at least for a time. If so, he reckoned without a sudden new intervention on 12 December from Haviland after some weeks of silence, in a memorandum to the RSA which it passed on to the LGB on 22 December. Haviland seems to have had no knowledge of the workhouse question, but he was still concerned about the sanatorium and the proximity of its pits to the water company's new site. Haviland stated that he had just revisited Uppingham, and believed that the issue of the waterworks site was so urgent that he was writing to the LGB on the move 'while travelling between Uppingham and Oundle', rather than leaving it all until he got back to Northampton. He complained bitterly about the location near the sanatorium. The old sanatorium cesspits had not been removed from the area, which had been a key place in the troubles of a year earlier. The water company's new well did not seem to go deep enough and was far too close to the old pits. Haviland went back over all the scarlet fever cases earlier in the decade. The company had 'signally failed' to provide an acceptable source, and he would not answer for the consequences.[32] He favoured an alternative site further to the north-west.

It is not clear why Haviland suddenly wished to revive all this controversy again,[33] but his intervention served to draw more attention to Candler's question about the workhouse. On 14 December Bell wrote again to Candler. He had kept up his pressure on Barnard Smith to do away with the eight or nine workhouse cesspits – both because of the health hazard they represented to the users of the privies, and also because of their proximity to the nearly-completed waterworks. But Barnard Smith was fiercely opposed to spending yet more ratepayers' money on the abolition of the pits and the provision of mains-connected water closets there, and he was supported by Wales. Bell received very little support from other RSA members, even though he had warned them that it risked being accused of double standards if it did not act speedily to remove the workhouse cesspits. There was another acrimonious debate within the RSA. Barnard Smith said that it was not possible to spend money on installing water closets in the workhouse. Bell countered that it had a well on the premises, so there should be few problems. Barnard Smith did not see 'why we should go to the expense of filling our cesspits to please the water company': he did not wish to boost its income further. Legally the RSA could not be forced to act – although when Barnard Smith tried to act unilaterally to close the issue down, Brown advised him that debate should be allowed on the question.[34] Whether prematurely or not, the *Stamford Mercury* expressed pleasure that the pits might be removed as soon as the new water and sewerage improvements were complete.[35] The *Lancet* printed a letter from 'A guardian', expressing a similar hope.[36]

News about the progress of the water company was mixed. On the one hand, construction work was now gathering pace on its site near the sanatorium. The first shares were being taken up. In a letter to J. C. Guy (whose bank was acting for the company), Thring subscribed £30,[37] and Guy issued a general invitation to the masters at Borth to follow suit. Christian was one of those who responded – buying 11 shares at £1 each.[38] But it seems that demand for shares was low amongst townspeople in Uppingham, partly because some people resented a Thring-led enterprise and also because the financial pressures caused by the school's absence were increasingly being felt. Thring wrote to Guy on 30 September:

> I do not understand the people of Uppingham. I fear I never shall. How people with property in the town can calmly run the risk of seeing it destroyed in value for want of drainage and water supply, and how people with hearts can be indifferent to the illness and death of their neighbours is beyond me. However fortunately I am not required to account for this.[39]

He was also keen that Guy should become a director of the water company, and had suggested this for some weeks, believing that Guy's dual roles as the bank manager and clerk to the school trustees made him well qualified for it. Thring and the masters would support Guy, because he shared their view that this was a venture for 'a community of which we are all members' rather than an issue of school versus town.[40] Thring may also have had in mind that fact that Brown had now begun yet another bout of acrimony in the letters column of *The Times* on the RSA's behalf with Hodding and Beevor, who were still acting for the water company. The solicitors had taken exception to some of the remarks in one of the letters to the paper from 'An old inhabitant', which had implied that the company was entirely a school initiative designed to thwart the RSA's own attempts to provide a mains water supply. They pointed out that a public meeting of townspeople had backed the company, and that a number of them were amongst its leading proponents. Every effort had been made to conform to the wishes both of potential consumers and of the RSA itself. They criticized the RSA's failure to provide a water supply in earlier years, and its insistence earlier in 1876 that no street should be dug up without its consent: 'a stipulation which would practically have rendered the act [i.e. the water bill] a dead letter'. They concluded that the slow progress of the project was due not to a 'want of energy' on the part of the company, but to the RSA's own attempts to stop the bill in Parliament – a move which had also incurred high costs. 'Peace had to be purchased from the sanitary authority by payment of £75 in aid of their costs of opposition, this being preferable to the expense of a parliamentary contest'.[41] Bell seized on this as another stick with which to beat Barnard Smith at the next RSA meeting.[42]

With the prospect of the school being back in Uppingham in January receding, Smart had applied to the RSA at the end of November for an extension of the time in which to complete the sanitary works. This had been granted; the RSA had little alternative. Bell believed that in drawing up the tender, it had grossly underestimated the amount of work involved.[43] But there was worse news to come. In late November and early December, there were yet again a few more typhoid cases in the town. Thring had already been receiving rumours for some weeks of this, and he wrote on 5 December to alert the LGB to the growing likelihood that the school would be unable to return at New Year, adding that the LGB had never fixed a new date for Dr Power's postponed inspection. He recommended a Dr Buchanan. The LGB took great exception to his making such a recommendation, replying that the school must decide for itself where it should reassemble in January; such questions did not fall within the recent Public Health Act.[44]

Hodgkinson, whose recruitment to the Lower School was now feeling the effects of the school's prolonged absence, wrote that it was 'very disastrous to me that the school [is] not returning'.[45] The reaction was even gloomier in Borth. Although Thring had been able to hide the implications from the boys, the masters understood all too well what renewed typhoid would mean. Coming at the end of a long hard term the news from Uppingham lowered morale dramatically. Many of the masters did not want to spend Christmas away from Uppingham, and they had left Borth as soon as term ended, even before any decision had been made as to the date of the school's return to Rutland. However, although staff understandably wanted definite news, it made sense for Thring to keep all options open as long as possible. He was secretly resigned to the fact that they would be back in Borth in January – and it may be revealing about his reading of the state of mind of both staff and pupils that he chose not to tell them before the end of the autumn term.

On one issue at least, Dr Bell in Uppingham and Dr Childs in Borth were united. Even though they might revive the ire of the RSA if they brought in another national expert to give a view on whether it was safe for the school to return, it must be done. Thring agreed, and Professor Acland, Regius Professor of Medicine at Oxford, was contacted.[46] Acland visited Uppingham on 18 December. His visit was very thorough: armed with copies of the reports by Haviland and Rawlinson, he toured both the town and the school. He met Bell and Childs, as well as Tarbotton, who had advised the school on building improvements a year earlier, and he called on Hodgkinson and Wales. Bagshawe and Rawnsley returned from Borth to put the housemasters' view, and Acland subsequently received a visit in Oxford from Haviland and Childs.[47] He was firmly persuaded that it was not yet safe for the boys to return, although he was confident that it would be so when the work was eventually complete.[48] Jacob came down from Liverpool, anticipating that he would need to brief the trustees and that Thring would need support against them. They met on 22 December. Faced with Acland's report, they had no alternative but to agree to the school remaining away for a further term. They also decided that they could come to no final settlement of the year's accounts until Thring sent them more details. They voted a further £300 to pay masters' salaries and £250 to Thring towards his expenses, but asked him to provide them with a statement 'showing in detail the value of the property belonging to the masters conjointly and separately for which they consider themselves to be entitled to be indemnified under the scheme'.[49] They were apparently at last giving some thought to the longer-term financial implications of the situation for the school and its employees.

Only two days before this, Thring had written to Christian, who had returned to Uppingham some days earlier to play a major role in arranging Acland's visit. He thanked him for this, but could not hide his weariness and dejection. The masters still with him in Borth were disputing financial matters:

> I am glad that you are cheered. I should be if I were not so tired, and so worried. Campbell has wasted my time this last fortnight by refusing to pay his scholarship contribution, for the life of me I cannot see a particle of sense in his letters. He has now paid. Rowe [who had left the school for Tonbridge a year earlier] has opened up on me about the capitation fees. I have referred him on to Tuck [the master who was acting as bursar] ... I shall want a secretary for the next three months and a lawyer at the end. My letters are such a heap ... I write from 10 to 1 daily without stopping, and the inside of my head feels as if I was growing a fleece there. Nevertheless I think, if I could think, that there really is a break in the clouds, and some glimpses of light under them ...[50]

Thring stayed in Borth for Christmas. There was little respite for him. Even on Boxing Day he was at work – writing to the parents to announce one more term in Borth, and assuring them that in no way would the efficiency of the school suffer.[51] Even the normally highly supportive John Skrine criticized him for the strained relationships of this time.[52]

The final days of the old year brought a wholly unexpected new development. Edward Wortley had taken the chair at the RSA meeting on 27 December, because Barnard Smith was absent – something almost without precedent.[53] Barnard Smith's death, from typhoid, occurred at Glaston Rectory two days later. It is hard to know how whether the burdens of running the RSA over the previous few months had contributed to his death. His attitude to the final dispute in which he was involved – his dogged and obdurate opposition to the abolition of the workhouse cesspits – perhaps suggests at the end a man on the brink of exhaustion and losing his sense of perspective. Whilst protecting the ratepayers from further expense was important and no doubt loomed large in his mind, it seems surprising that he did not grasp the potential effect on public opinion of his opposition to improvements in the RSA's own workhouse.

Even in an age accustomed to sudden death, Barnard Smith's passing caused deep shock. The news reached Borth on New Year's Day. Thring's reaction in his diary was regretful but unyielding:

> The sad and fearful news reached us that Barnard Smith has died of typhoid fever – apoplexy the immediate cause. Poor fellow! He has fallen a victim to his own obstinacy and delusions. It brings home to us very close, 'He forgives our trespasses as we forgive'.[54]

He wrote to Bell:

> Your news is truly awful. I am very, very sorry for him. God is very gracious and hearts are not open for us to read, yet nevertheless it is fearful to be suddenly taken away whilst doing wrong. God help us all.[55]

The RSA members met again on 3 January 1877 and formally recorded the 'unexpected and deeply lamented death of the Reverend Barnard Smith. They cannot refrain from placing on record their strong and grateful sense of the services he has rendered the board while acting as its chairman this last nine years.'[56] The *Stamford Mercury* described him as being noted:

> not only for his literary labours, but also as a staunch friend to educational pursuits, and also a most successful tutor in the University of Cambridge ... he devoted his talents ands experience for the benefit of the ratepayers within the union. There was not a charity or valuable institution within the neighbourhood of which he was not an active member. His loss will not fully be recognized until time shows the actual value ...[57]

However, with Barnard Smith gone, there was an opportunity for a fresh start. Bell was keen that Wortley, who had written in such conciliatory tones only a few weeks earlier, should become permanent chairman of both the RSA and the guardians. In the event Wortley took on the RSA chairmanship, but Wales, perhaps feeling that Barnard Smith's death was due in part to exhaustion, suggested that the combined role was now too onerous, and the Hon. W. C. Evans-Freke became union chairman.[58] Bell declared himself content with this. He was, however, still determined to pursue the question of the workhouse cesspits and immediately demanded to know what legal powers the guardians had, or needed, to make structural alterations and how they might be paid for. This would mean more dealings with the LGB.[59] In less troubled times the RSA members might well have tried to block a move which implied more work and additional expense. However, shocked by Barnard Smith's sudden death, they were having second thoughts about their previous indifference to the workhouse question. They agreed to ask Rogers Field to draw up the necessary designs for a new system. Copies were sent to Haviland and to Dr Walford, who was medical officer to the workhouse.[60] But even this apparently uncontroversial decision produced a new burst of acrimony – this time between Haviland and Dr Walford. Walford, supported by Bell, was very keen on linking the workhouse to the new sewerage system, with water closets to be flushed with water from the water company's pipes. Haviland was adamant that dry-earth arrangements (of which he had long been a supporter, unlike Rogers Field)[61] and would be a better solution than water closets, given the local gradients and soil conditions. He still claimed that the water pressure might not be adequate for water closets, as the workhouse was on some of the highest ground in the whole town. Wortley, who had used a dry-earth system successfully at his own property for many years, was inclined to agree.[62] Bell used the controversy as a further opportunity to condemn Haviland for involving himself in the workhouse issue only at such a late stage:

> It is a most extraordinary circumstance to me that Haviland never found out that there were cesspits at the union ... If he knew of them he kept them very dark, and I think his opposition arose from his annoyance at my having brought them to light.[63]

The bemused RSA members appealed to Rogers Field to give advice.[64] The LGB was keen to avoid involvement in the question; it had already decided not to involve itself in the dispute about the proximity of the sanatorium cesspits to the new water station, and it believed that it was far too late to start querying the water company's arrangements at this stage. In an internal memorandum on 3 January it recorded: 'There does not appear to be reason for distrusting [Haviland's] conclusion ... [but] the report seems to be sent for the [LGB's] information, not in order that the [LGB] should express any opinion on it'.[65] It endorsed Haviland's suggestions about dry-earth arrangements at the workhouse, and the alterations went forward, despite Bell continuing to lobby Evans-Freke.[66] The RSA confirmed this decision on 21 February.[67] The inhabitants of the workhouse would not yet enjoy the same facilities as other inhabitants of the town.

The meeting on 21 February was another robust one, with members divided as to whether to seek to involve the LGB further, and whether or not to respond to another recent attack by Haviland on the water company. In the end it was decided that it was too late to oppose it anyway; they would let the issue drop. On the same day Bell reported that 'the flushing cart has arrived, and the sewers are being swept out with it'.[68] He had already been in touch with Christian at Borth to report that 'the health of the town is good, very little illness indeed', although he had recently seen two child cases of typhoid caused again, he believed, by polluted wells.[69] Nevertheless, the overall picture was optimistic, and supplies of water would soon start to flow. The *Stamford Mercury* had reported on 9 February that preparations were being made for the school to return after Easter. One worry remained, however; Haviland's continuing hostile preoccupation with the water company – despite the fact that its share capital had been fully subscribed, construction work was well advanced, and mains pipes had been laid along every street and up to all the houses.[70]

Meanwhile down at Borth the storms at the start of the new term had given the school a further rude awakening to the realities of Atlantic coastal life, and a reminder of the urgent need for progress in Uppingham.[71] It would also be important for the school to have plenty of supporters within the RSA after its return. Its annual spring elections were not far away. It was unlikely that the masters would have to vote from Borth station platform once more, as they would surely be back in Uppingham after Easter, but Bell felt the need to be prepared, just in case the RSA tried underhand methods again. He was worried that the rector might be hoping that there would be enough candidates to edge out Bell or some other of the school's supporters. On 20 February Bell wrote to Mullins, who appears to have been the spokesman for the masters as a whole, about this question. Should they try to avoid a contest by only putting up Bell himself, or should they also nominate another ally of the school? Would it matter anyway, once the sewerage works were complete and the water company was in business? He was sure that they were not out of the woods yet:

> The animosity is not dead, Haviland has been showing his teeth, and some of the Authority will back him the moment the year of grace [for the water company to complete its work] expires, as at present they feel powerless to do any serious damage to it.[72]

On the eve of the elections Bell even tried to persuade Thring himself to stand against the rector (following a further dispute about which we have no details). For Thring – who had recently written to a friend that only Professor Acland's visit had prevented the 'Rutland clique' from forcing him to bring the school back to Uppingham at Christmas or to resign instead[73] – the prospect must indeed have been tempting. However, in the final days of exile, this would have been one battle too far. In almost the last letter he penned from Borth, Thring replied to Bell:

> I had heard what an astonishing exhibition the rector made of himself at the meeting. This last year seems to have taken him quite out of his depth, and upset all his shallow water experience ... If the form the thing takes is <u>starting</u> a candidate in opposition to the rector, I do not think I ought to let myself be nominated, but if the rector is ousted first, and then there follows an election, I should not object to take it. I see no objection whatever to you standing ... it is absolutely necessary for the school to ... take part in the parish politics. But I could not bring myself to appear before an Uppingham audience as in any way challenging direct comparison with the Rector. He is no antagonist for me.[74]

Pateman, the school's second choice candidate after Bell, initially intended to decline to stand again, but changed his mind in anger when he heard about Haviland's continuing opposition to the company.[75] In the event, he and Bell were elected.[76]

Bell had one further, personal issue to revive. He had heard periodic gossip which had filtered back from Borth about the widespread praise there for Dr Childs. It would culminate in Child's being afforded a hero's farewell there,[77] and Bell was concerned to ensure that Childs should not be allowed to continue as the school's medical officer once it returned to Uppingham. He wanted to protect his own position and his income, so it was hardly surprising that he suggested Childs would not have the time to do both teaching and doctoring. Bell was convinced that Thring had promised, back at the end of 1875, that Childs would be employed merely as a science master. Childs, however, was determined to continue practising medicine at least to some extent, and claimed that Thring now proposed that each housemaster should choose between the two of them. Bell feared that if this went ahead, Childs might later resign as science master and start a practice of his own in Uppingham. He was 'in doubt as to whether he should trouble Mr Thring' about it; it seems clear that he hoped Jacob 'would do the troubling for him'.[78] Jacob replied that Bell should write to Thring himself, which Bell duly did on 21 March.[79] As it turned out, he need not have worried. Thring wrote back confirming that, on this issue, Bell would have his support.[80] Bell wrote at least two letters of thanks in response. He also suggested that it would not be wise for Childs to publish a report on the typhoid outbreak which he was apparently planning. Such a report might not suit Bell's own purposes, given some of the earlier accusations of negligence made against him, but he was also correct to reckon that, with a surfeit of reports already, a further one would merely revive old controversies. He wrote to the ever-patient Jacob a number of times on the issue, culminating in a long self-justification on 21 March, following a visit from Childs himself.[81]

The *Stamford Mercury* confirmed at the start of April that the school's return was fixed for 6 May.[82] The works were done; water was flowing and new drains were in place. The trustees had directed – this time uncontroversially – that the school should return at the start of the following term. But they were now in dispute with Thring over the size of his claim for expenses, and they decided to require that he 'state each term what sums be required for plant and apparatus, and that he make special application to the trustees before incurring any expenditure beyond the amount granted for this purpose'. Once the school was back, they were determined to tighten their grip on its administration – and its headmaster.[83] The return would not be a moment too soon, either for school or town. For the school, it had been a hard winter. For the town, times were tough too. There appear to have been only one or two bankrupt businesses in Uppingham during the school's absence,[84] but it is likely that the overall economic effects had been marked. The *Stamford Mercury* carried its annual description of the spring fair in Uppingham. This year, despite 'the usual accompaniment of steam-horses, swing boats and rifle galleries etc ... not much business was done'.[85] With the school absent for so long, money was tight.

The school was still in session at Borth over Easter. After a farewell concert in Aberystwyth, and a lengthy and effusive farewell celebration in Borth to which almost the entire village turned out, the pupils were gone.[86] 'And so the grand page of life is turned', wrote Thring on 13 April, 'the chapter come to an end. But it has been glorious'.[87] He returned to Uppingham on 24 April 1877:

> with wonderfully mixed feelings ... thankfulness to God for a page turned and closed; intense dislike of the place, mixed with a feeling of home and being master once more in my own house; the old constriction of stomach and feeling of dread, mixed with a sense of no longer being at the mercy of others, and subject to the racket and disturbance of hotel life ...[88]

Messages of congratulations on the school's return poured in – including one from a fellow headmaster: 'In my judgement your exodus was one of the bravest exploits ever performed, and you deserve to be hung all over with Victoria crosses ... you will be in the world of immortal achievements'.[89] A week later Thring noted:

> The town is really making a grand demonstration: arches and flags all up in the street, and they must have taken much time and care and spent much money in doing it. This calling out of feeling and drawing attention to the school, is a new start in life here ... a signal refutation of the calumnies vented on us last year, and the whole moral atmosphere of the place will no doubt be changed in future ...[90]

There were banners and triumphal arches of evergreen: 'Welcome home', 'Flour-ish school: flourish town' and 'Uppingham School: a good name lives for ever'. These heralded two evenings of triumphant processions soon after the pupils returned. 'The reception on Saturday night was even better than Friday, and the whole town was in a wonderful fervour of enthusiasm'.[91] The *Stamford Mercury* recorded it all in great detail, praising Thring's 'determined efforts', and describ-ing how flags were hung from houses, with so many streamers and so much bunting 'that it would have done honour to a royal visit to a town four times as large as Uppingham ... There was scarcely a house which did not contribute its quota towards the gaiety of the scene'.[92] Mr White, the heroic carrier of the voting slips a year earlier, had large welcoming displays outside his ironmonger's shop in the High Street – as did Mr Dolby, the tailor a few doors away. Dr Bell's surgery was bedecked with Chinese lanterns.[93] The bus from Seaton arrived; its horses were detached and pupils then dragged it around the town. Bands played, and many cheers were given. The only sour note was sounded by Wales, who ini-tially declined to have the church bells rung – possibly out of pique that Thring had recently been elected to replace him as president of the town-school mutual improvement society.[94] In the end, even he sensed the celebratory mood and reversed his decision.

Three days later there was a ceremony at the school itself in which speeches of welcome were given by Dr Bell and by John Hawthorn, who had played such a key role in the ratepayers' revolt in the previous August. Hawthorn would have had as much reason as anyone to welcome it back, and he observed that 'the absence of the school had pressed with severity on many tradesmen'.[95] Thring was presented with an illuminated address, and replied at length: 'We are united now as we never have been before'. He also remarked that with the summer intake of pupils in addition to the sixty-six who had joined the school whilst it was at Borth, nearly a hundred boys were seeing it in situ for the first time. The parents had remained loyal despite their criticisms when typhoid had first struck.

Normal life quickly resumed. In an effort to maintain the new spirit of coop-eration, a town-school feast was held later in the summer, and a joint cricket match took place against a Derbyshire eleven. A new recreation committee was planned; among its first achievements were a flower show, a concert and an ath-letics gathering, as well as a big Guy Fawkes night celebration.[96] Lectures on a variety of topics continued through the winter, as well as cookery and elocution classes and a number of play readings.[97]

An increasing number of households were now being linked up with the new sewers and the new system appears to have had the desired effect, although there were continuing calls for the abolition of all cesspits.[98] Even so, Upping-ham would not be disease-free for some years. There was a brief scare in late November 1877 when scarlet fever was reported at Bagshawe's boarding house up on the hill,[99] but fortunately the case proved to be an isolated and mild one. Three smallpox cases were recorded in the town three months later, one of which proved to be fatal.[100] Late in 1878 a small-scale typhoid outbreak caused a new scare about possible impurity in the water. Brown, who was still RSA clerk at that date,[101] was bitterly criticized by Bell and others for causing alarmist rumours to fly, by acting too slowly in getting a water analysis carried out. This eventually proved that the company, which was now providing its mains water service, was not to blame.[102] In many respects it seemed to be performing well; it had reached an agreement with the boarding housemasters for them to enjoy reduced charges in recognition of the fact that their pupils were in Uppingham for only part of each year.[103] The LGB had approved its regulations. By June 1880 it was working with the RSA to adopt a hydrant system for extinguishing fires, flushing drains and watering the streets. The *Stamford Mercury* reported that 'the water company agreed to put at the authority's disposal their tank of 30,000 gallons, and by starting their pump supply, 5000 gallons an hour could be kept up'.[104]

The company later ran into trouble, however, justifying the earlier fears of both Haviland and the RSA about the adequacy of its specifications. The new wells which had been sunk between the sanatorium and the workhouse initially produced large quantities of water – to the extent that the whole site around the new water tower became flooded. But the water table soon dropped and the supply became insufficient as demand increased.[105] Two years later (1882) the summer supply was restricted to less than an hour per day, and in December 1883 in a desperate attempt to find additional supplies, the water company sank a large well costing £500 to a depth of 112 feet. It found nothing.[106] Headings were then driven from the bottom of the existing well in various directions before a new supply was discovered further to the north, which solved the problem for a while,[107] and there was sufficient water in August 1888 for 'the old bathing place on the Seaton Road [to be] filled with water, after having been empty for several years'. Boating was provided on the August bank holiday, with a band, dancing and fireworks.[108] In same year, however, when one new boy arrived:

> ... a school and town water-supply that was unscientific and somewhat precarious not infrequently gave rise to the rumour that if it did not rain we should be sent home, and supplied the perennial jest retailed to newcomers that the water in the school bath got so thick by half-term that once an adventurous fag, adept at diving and of name unknown, had in some past era, also unspecified, dislocated his neck by diving into the mud.[109]

Despite the problems which the company faced in these years, by 1900 its shares were selling at £6: six times their 1876 price.[110]

As in earlier periods in this crisis, in the final phase it was local rather than national personalities, opinions and events which determined the various outcomes. Thring's action in removing the school and Barnard Smith's untimely death probably did more to achieve a unity of purpose in Uppingham than any other events. They also ensured that the prevalence of the trader interest over that of the ratepayer/landowner would not be reversed – even though the rate increases would prove formidable. It is very unlikely that acceptance of such a rapid increase in sanitary spending by a small RSA could have been achieved without these exceptional circumstances. In normal times, guardians who suggested such radical improvements would surely have been turned out at the next local election. The extent of the crisis also made it far easier for the private water company to win local acceptance.

The origin of the Uppingham typhoid outbreak and the identity of its carriers were never established, yet the battles surrounding it played a distinctive part – not only in securing proper sanitary and water improvements for the town of Uppingham, but also in the development of medical care in boarding schools. This can be seen in the successive, ever-larger editions of Clement Dukes's book *Health at School*. His 1905 edition laid down the key ways in which the risks of epidemics could be minimized: instant isolation of the first case; rigorous quarantine arrangements thorough disinfection, and the burning of all books used by the infected person; plenty of space and ventilation in dormitories; efficient drainage; pure water; and a high state of pre-existing health in pupils to build up resistance to infection.[111] Whether Dr Bell ever read Dukes's work, and whether, if he did, he heard echoes of Haviland within it, is not recorded. In the years after the school's return he renewed his complaint that Haviland had no right to visit his patients, with an added protest that Childs was also seeing some of them: both pupils and local families.[112] After their time, as John Honey records:

> In the early decades of the [twentieth] century, a schoolmaster could still notice that illness was common enough to be a major topic of conversation in public schools – 'what epidemic sickness had plagued the school last year, or last term, and what was likely to plague this term'. Epidemics themselves were to become less common, and certainly less virulent, after the development of chemotherapy (e.g. M & B) in 1936 and the antibiotics in the 1940s, leaving the empty school sanatoria as huge white elephants to be adapted where possible in our own day as additional boarding houses.[113]

AFTERMATH AND CONCLUSION

Once life in Uppingham returned to normal, both town and school faced a financial reckoning. Where the town was concerned, the parliamentary local taxation returns for Uppingham and its immediate neighbouring towns (shown in full in Appendix 2) reveal the extent to which the RSA strained itself and its local community in an effort to effect the necessary improvements. Barnard Smith and Wales, who had warned so repeatedly about the extent of the burdens of sanitary reform which would fall on hard-pressed ratepayers, proved to be right.

Table 2: Money Spent by Uppingham and its Neighbouring RSAs, 1874–83 (in £).[1]

Year	Uppingham	Oakham	Market Harborough	Melton Mowbray	Stamford
1874	2,351	172	444	275	271
1875	639	81	775	570	288
1876	963	185	628	482	280
1877	3,227	371	514	500	285
1878	2,125	515	1,085	513	289
1879	1,489	2,135	789	455	274
1880	898	831	893	529	257
1881	1,175	659	673	226	1,074
1882	1,118	746	688	536	459
1883	1097	648	702	276	391
Total	15,082	6,343	7,191	4,362	3,868

Whichever way one views the statistics, Uppingham was under far more financial pressure in these years than the communities around it. Moreover, the loan which it was struggling to repay in the late 1870s was at a level exceeded in only twenty or so RSAs throughout the whole of England and Wales, most of which were markedly larger in terms of rateable value.[2] Uppingham RSA's spending on both sewerage and water provision compares well with its immediate counterparts, and the acceleration in its activity during 1874–83 compares very favourably with, for example, the almost static picture in Stamford.[3] Even allowing for year-to-year fluctuations caused by the dates on which each RSA paid its bills, and for the differing levels of spending by each RSA *before* 1875, it is clear that the property owners of Uppingham had to pay dearly for their improved

Table 3: Population of Uppingham and Neighbouring RSAs, 1871–82.

	Uppingham	Oakham	Market Harborough	Melton Mowbray	Stamford
RSA population 1871	12,443	11,142	16,081	19,926	17,821
RSA population 1881	12,029	10,978	16,285	20,483	18,334
Mean population	12,236	11,060	16,183	20,205	18,078
Town population 1871	2,464	2,911	2,362	5,011	6,686
Town population 1881	2,549	3,227	2,669	6,347	8,733
Mean population	2,507	3,069	2,516	5,679	7,709
Inhabited houses 1871	438	622	505	1,015	1,419
Inhabited houses 1881	432	675	544	1,263	1,847
Mean total	435	649	525	1,139	1,633
Rateable value 1875	97,015	92,658	147,173	136,359	91,556
Rateable value 1882	110,996	108,049	137,618	160,145	101,113
Mean total	104,006	100,354	142,396	148,252	96,335

Table 4: Spending per Head of Population and per Household by Uppingham and its Neighbouring RSAs, 1875–82 (in £).

	Uppingham	Oakham	Market Harborough	Melton Mowbray	Stamford
Spending ÷ mean RSA population	1.230	0.520	0.440	0.220	0.210
Spending ÷ mean town population	6.020	2.072	2.860	0.770	0.500
Spending ÷ mean total of inhabited houses	34.700	9.770	13.700	3.830	2.370
Spending ÷ mean RSA rateable value	0.145	0.063	0.051	0.029	0.040

facilities – and for the loans taken out to finance them.[4] Meanwhile the school faced even greater pressures. Thring had no doubt known that the costs would be substantial, but his bitter resentment of the RSA's long previous inaction had always prevailed over that realization. He and the housemasters were hit twice, because as householders in the town: they could not escape the costs of their new water supply and extended sewers. They also had to face personal financial consequences and issues from the decision to migrate. As he had feared throughout his time at Borth, Thring found himself deeply in debt.[5] Forced to end such luxuries as his annual expedition to the Lake District,[6] he appealed to the trustees for help. They showed scant sympathy and played for time, merely agreeing in the first instance (June 1877) to reimburse the travel costs of the day boys to Borth.[7] At their October meeting they passed a motion implicitly critical of Thring for a failure of accounting procedures, and at their next meeting, in February 1878, they criticized the level of spending on concerts and musical instruments. Their minute book also records that:

They had before them this day a memorial from the masters concerning expenses of the school at Borth. They find themselves without accurate knowledge of the amount and particulars of the expenses neither do they know who are liable for them, whether the masters as a body or the masters individually in varying proportions. They resolve to form a committee of investigation and request to be furnished with full information, when they will further consider the subject.[8]

This committee was chaired by Sir John Fludyer; Thring must have welcomed the inclusion of both Birley and Jacob, but Wales was also a member of it. The trustees came to believe that the debt could be gradually reduced by an increase in the number of boarders – something which they must have known Thring would strongly oppose.[9] At their April 1878 meeting they had also passed a resolution 'to bring the whole financial condition of the School before the Charity Commissioners'. The argument dragged on for some months and through several board meetings, before in October 1878 they agreed to grant limited payments to Thring and a long list of masters, totalling only £3,275.[10]

Thring had meanwhile contacted the Commissioners on his own account. He sent a petition on 15 April 1878, in which he urged that the Borth expenses should fall on the trustees. This stated that he 'approached the question with great diffidence', but he wrote with passion about how the school had been built up through the financial contributions which he and the masters had made, describing himself and his colleagues as 'the living representatives of the new foundation'. He suggested that the penalty to them of the autumn 1875 epidemic *alone* totalled nearly £4,000 in lost fees, and pointed out that hard on the heels of this had come the cost of the improvements to the houses. After the March 1876 outbreak there had been the additional expense of the move to Borth, whose costs he estimated at over £3,000, to which the trustees had contributed a mere £250 – only a third of what had been raised through Captain Withington's fighting fund. Another £300 had been given 'from within the school itself'.[11] He tried to show that the houses could not increase their boarder capacity. He also suggested that the Borth migration had merely exacerbated a long-standing problem: 'the impossibility of carrying on the school [under the fee arrangements fixed by the Commissioners a decade earlier] without an increase of funds'. He believed that an additional sum of at least £20,000 needed to be invested in plant and equipment if Uppingham was to function properly. His conclusion was that the tuition fee needed to be raised from £30 to £40.

The trustees, fearful at the financial consequences if the Commissioners were to back Thring's petition, tried again to evade all responsibility for the move to Borth. Thring wrote to the Commissioners once more on 24 May, protesting at this. He again described the sequence of events in 1875–6 and the immense pressures which he and the housemasters had faced. He sought to demonstrate that he had informed and consulted with the trustees at every stage.[12] His efforts were successful; the Commissioners were in no doubt that:

> ... although the removal of the school to Borth had not the express sanction of the trustees, yet their subsequent acquiescence in it must be assumed ... from the part they took in the management of the school during the time of its stay at Borth.[13]

They agreed to the suggested fee increase, but they exempted the small number of day pupils from this additional charge. The additional revenue would ease Thring's burdens, but no more than that. It seems certain that he and his colleagues never recouped much of the money expended during the move to Borth. The trustees did, however, agree to take over the sanatorium in 1878, together with its mortgage, half of which still remained unpaid.[14] The Commissioners added one further recommendation: that in the longer term the school should buy the houses from the masters. This was implemented eventually, but not yet. In the years after the Great War of 1914–18 the school began buying houses from their owners so that new housemasters would not have to bear the burden of purchasing them from their predecessors.[15] From 1946 it steadily ended the arrangement whereby they drew profits as boarding-house keepers. Henceforth housemasters would be paid a salary instead.[16]

In most non-financial respects, Thring had won the day. The majority of his housemasters remained at Uppingham until retirement, although his relationship with Hodgkinson, once so close, never recovered from the pressures to which the typhoid outbreak exposed it.[17] George Mullins, whose son had been one of the early victims of the typhoid outbreak, lost another son to pneumonia in 1893.[18] Thring's final decade as headmaster was an altogether quieter and more mellow period, as his fame became more widely known. He felt rejuvenated by his teaching.[19] Where the staff was concerned: 'One moves amongst the masters so secure and at ease and not on the watch any more for the next plot or stab'. The Borth commemoration on St Barnabas' Day each June became an annual event in the life of the school chapel.[20] He spent part of the late summer at Borth at least three times in the next five years, always warmly welcomed,[21] and in 1880 he was even greeted at the station by a brass band. He wrote to Christian in 1881: 'You will laugh when I tell you that I have been preaching at an Eisteddfod in Borth today'.[22] Financial concerns dogged him, however, for the rest of his life. He thought of retiring, but was still too concerned about how little capital he had been able to accumulate over the years.[23] He died, still in office in October 1887, aged 66 – a comparatively poor man; when his wife Marie followed, his five children would inherit barely £500 between them.[24] *The Times* recorded: 'a throng of mourners from all parts of the country'[25] at his burial in Uppingham churchyard, conducted by George Christian. One of the wreaths at his funeral came from 'the women of Borth'.[26]

Bagshawe's move to the Lower School in 1882 (in succession to Hodgkinson) aroused new concerns for Dr Bell, who had long known that Bagshawe was no admirer of his. Bell sought the trustees' assurance that even if they had no formal power over the Lower School, they would use their influence to see that Bagshawe did not dispense with Bell's services for his pupils.[27] Bell remained in general practice in Uppingham, and eventually became medical officer of the workhouse and public vaccinator too, on the retirement of Dr Walford.[28] He also contributed one article to the *Lancet,* in 1899, entitled 'A Woman Disembowelled by a Cow'. For many years he was a JP and churchwarden. He died on 11 July 1914. The *USM* in its tribute understandably ignored the pricklier side of his character and reflected on all that the school owed him for his steadfastness during the crisis years:

> His life was a constant influence for good, both in school and in town ... The fact that he would not give up work, and was, within a few days of his death, attending some patients, is a striking example of that persistence ... keeping duty strictly in sight throughout. Who shall say that England does not need such lives?[29]

Less than a decade after the events which threatened Dr Bell's career, in 1884, the Medical Officers of Schools' Association was founded. One of its first tasks was to draw up guidelines for guarding schools 'from the outbreak and spread of preventable infectious diseases'.[30] Three years after its inception, Clement Dukes published his first edition of *Health at School.*

Bell's arch-enemy, Dr Alfred Haviland, retired as MOH for the Northampton combined districts in the early 1880s, and went to live for fifteen years on the Isle of Man, first in Peel and later in Douglas.[31] He threw himself into local life there, and was much in demand as a writer and lecturer, particularly about the island's climate, glaciation and geology.[32] Shortly after his arrival on the island he appears to have met his match as a controversialist in the shape of Revd Theophilus Talbot, as a result of two lectures in 1883.[33] In these, Haviland praised the healthy Manx climate, and suggested that it resulted in a very low numbers of cases of consumption there. These provoked a furious response in no fewer than five long review articles by Talbot, a local antiquarian, who claimed that Haviland's research was hasty and superficial, that he grossly underestimated the rigours of the Manx climate, and that incidences of consumptive illness were far greater than he had been able to understand, during his as yet very brief time there. Talbot also attacked Haviland for trying to raise funds for the publication of the lectures.[34] His comments are perhaps significant in view of the bitter criticisms which Haviland's research methods and judgements on Uppingham had provoked from Thring and his supporters, less than a decade earlier. Thereafter Haviland compiled a 21-page pamphlet on Port St Mary as a healthy resort, and a series of recommendations about new house buildings and drainage.[35] He played a leading role in the inauguration and continuation of three island socie-ties, devoted to astronomy, antiquarian studies and medicine.[36] He also sketched and painted the local scenery, and returned to the mainland at the turn of the century.[37] He died in at Frimley Green, Surrey, in 1903.[38]

Revd William Wales retired as Rector of Uppingham only two years after the school's return, living first of all in Northamptonshire and finally in Leamington Spa. He died in 1889.[39] His steward (and RSA clerk) William H. Brown did not last long after Wales left, resigning in 1879 after it had been discovered that he had been stealing clients' money entrusted to him as a solicitor to invest.[40] J. C. Guy remained as clerk to the trustees until 1909. A year later, the school appointed its first bursar,[41] in recognition of the increasing financial complexity of running such enterprises.

Robert Rawlinson received a knighthood in 1883 and remained chief engi-neering inspector of the LGB until 1888. Rogers Field was back in Uppingham again in 1879, recommending further extension to the sewage farm on Seaton Lane.[42] His career included advising Wellington College on its diphtheria out-break[43] and designing the drainage systems for both Sandringham House and Bagshot Park. He used his Uppingham experience as the basis for a highly detailed handbook on sanitary bye-laws which was adopted for national use by the LGB in 1877.[44] He also invented a new type of aneroid barometer.[45] The LGB remained in existence for another forty years, although its relationship with local authorities was significantly changed by the setting-up of county councils and county boroughs under the Local Government Act of 1888. In 1918 it was reorganized and renamed the Ministry of Health.[46]

The Introduction to this book suggested that the Uppingham epidemic remains significant in three key areas: the inadequacies in local and central gov-ernment system at the time, the limitations of contemporary knowledge about epidemic disease in rural areas and confined communities, and in enabling us to assess the impact of local rivalries and personalities in such communities, the most notable of whom was Thring himself. It is worth returning to each of these themes in turn.

This investigation reveals the huge, growing gap between the expectations on local government and its officials in the 1870s and their ability to deliver sanitary improvement. It supports the view of the Webbs that only the state could bring about the degree of collective protection which the growing communities of the time needed and demanded, because the guardians had little prospect of being able to do so.[47] Limitations of time, capability, willingness to be involved and a variety of types of expertise made the work of rural guardians all but impossible by 1875 – at least, other than in peaceful and routine times. With growing pub-lic awareness of sanitary issues, and rapidly rising expectations about the delivery of improvements – especially in communities which had undergone rapid population increase – the period between the Public Health Acts of the early 1870s and the setting-up of county councils nearly two decades later was a time when the demands on RSAs became almost overwhelming, especially if exceptional cirucmstances arose. They really were caught between the mutually contradictory desires of local people for economy on the one hand and decisive action on the other. Uppingham shows that the rural 'shopocracy' was every bit as real as the contemporary urban one identified by Hamlin.

In routine matters of administration, the Uppingham RSA deserves a better reputation than it has previously enjoyed – for example in respect of vaccination and day-to-day workhouse management before 1875. But on issues of strategic planning and large-scale capital expenditure in an unexpected emergency, it was found grievously wanting. Throughout the two years 1875–7 and arguably for many years before that date, it struggled reactively to respond to the demands of the growing town, rather than being able to determine outcomes proactively.[48] It appears to have done little to prepare the ratepayers for the need for greater expenditure. It chose to ignore the impact of a growing number of pupils in a small town in term-time on both sanitation and water supply, and the risk from cesspits and polluted wells. It signally failed to ensure that those recent works which had been put in place were maintained and cleaned. It is debatable as to whether it could have done more to encourage local householders to link their houses up to the new drains. As a result, both sanitation and water supply had fallen well behind the town's needs. In its defence, there were no definitive guidelines about the levels of sanitation expected in each community, nor any national consensus on what sort of organization (public or private) should provide water. Moreover, the overwhelmingly agricultural nature of the area, and the massive landed interest represented both within the guardian body itself and the board of school trustees was completely incompatible with any radical and expensive reform. This too supports the Webbs' argument.[49]

Throughout the epidemic Thring derived his strongest support from the two trustees who were *not* local, and who had a business background rather than a squirearchical farming one. Birley and Jacob would have experienced first-hand the reforms already taking place in large cities such as Manchester and Birkenhead/Liverpool in the north-west in which they lived, and the desirability of good sanitation as a spur to economic and industrial growth. In that respect, they had much more in common with the urban local leaders identified by Garrard[50] than with their fellow trustees or the Uppingham guardians. The influence of a number of dominant school parents living in those cities served as a spur to action: the school knew that if it failed to be decisive, those parents could easily take their sons away to be educated elsewhere. Once the crisis broke, a mixture of fear, ineptitude, shortage of money and unfamiliarity with the detailed issues involved forced Barnard Smith and his colleagues into a defensive and procrastinating posture towards the school, sanitary reform and the private water company. As the pressures mounted, they allowed their hostility to Thring and their dislike of the way in which he had moved the school away from its local roots to harden into defensiveness and inactivity in the face of crisis. This failure of leadership may have stopped them preventing the spring 1876 outbreak, and it certainly caused the school to take itself off to Wales, initially for one term but eventually for a year. During the spring and summer months of 1876 they showed a remarkable lack of urgency.[51] They can be accused of double standards in their outspoken criticism of the school at the same time as they presided over sanitary conditions which were clearly inadequate in both the workhouse and the national school.[52] However, they faced a huge number of obstacles – including the ineffectiveness of overlapping local agencies in the years immediately before the passing of the Public Health Acts; limited legal powers; disputes with powerful landowners; a slow-to-act LGB; loan requirements and technical difficulties. They also faced demands for time and expertise which voluntary, amateur officials were wholly unequipped to meet. They had far fewer resources, both in terms of finance and of personnel, to call on than larger towns.[53] They could not be expected to be experts on the legal and technical complexities of sanitation and water supply. Their financial room for manoeuvre was limited partly by the small size and wealth of the geographical area they controlled, and their powers may well have been inadequate (as they often claimed, in their quest for USA status). Even if they had formulated a better improvement strategy and set aside the finance for it in the fifteen years up to 1875, they did not have Hennock's other two prerequisites for an RSA wishing to carry out a sustained programme of improvements: political skill, and substantial revenue over and above the rates.[54] In an age before the existence of county councils with their paid officials, RSAs themselves had to commission private consultancy in their search for technical advice. We can only speculate as to precisely what Rogers Field charged it for his work, but Appendix 2 suggests that it is likely to have been formidable compared with their expenditure in normal times.

The guardians had legitimate engineering and geological technical concerns, justified later by the town's continuing shortage of mains water in the years up to 1900. They could not afford the prospect of the costly digging-up of roads, an act over which it would have had no legal control. They feared that a private water company monopoly would quickly lead to rising water charges for local inhabitants. A shortfall in capital accumulation would explain their opposition early in 1877 to the speedy installation of water closets in the workhouse.[55] The same problem, together with sheer exhaustion and unwillingness to get involved in yet another legal and technical dispute, might explain their failure to give Haviland any practical support in his opposition to the new waterworks site in, the spring of 1877. The guardians and their successors would also have been the ones who would have continued to have to deal with the long-term financial and practical effects of any major sanitary improvements, long after Thring had died. The rapid rises in rates after 1877 show that their earlier caution was based on a sense of realism. They could not raise rates at will: the local ratepayers would and could take only so much by way of increased rates. The pressure of mortgage borrowing in Uppingham was heavy, as those RSA members who were lawyers would have been all too aware.[56] They were right to fear local resistance to improvement on both financial and practical grounds. Far from demanding urgent action early on, the traders were comparatively quiescent for twenty years before 1875, and then through three outbreaks of typhoid; even when the prospect of the school's departure seemed imminent in March 1876 they were at best half-hearted, leaving the housemasters to lead the opposition at the ratepayers' meeting. Only the prospect in August 1876 of the school staying away for another 4–8 months or even permanently, stirred them into decisive resistance to the RSA. It was shopkeepers Hawthorn and Compton who were prepared to give the eventual lead. This change of view, along with the supreme irony of Barnard Smith's death from typhoid, was the driving force which led to improvement in the end.

These issues challenge some of our modern assumptions about government infrastructure at national, and particularly at local, levels. By contrast with the 1870s, early twenty-first century governments employ an army of paid officials including national government task forces and mobile crisis teams to deal with unusual disasters or to enact complex legislative change on the ground, to offer sophisticated medical expertise and to provide economic protection for individuals at risk. These were evident in the government's mass-slaughter response to the outbreaks of foot-and-mouth disease in rural areas in 2001 and 2007.[57] In today's world such officialdom is largely accepted by the electorate – although there are perhaps some modern parallels with the hostility to Haviland as MOH in the criticisms made of council tax officials and health and safety inspectors. In the 1870s there was a much stronger belief in the proper limits of government and officialdom: arguably more laissez faire and less nanny state.

One can go further in the RSA's defence. The way in which both the town and the school populations had grown during the previous half-century, and the extent to which improvements had already been carried out, highlight the difficulty that the RSA had in balancing progress, cost and public opinion. They had no direct control over the school's growth, and no means of making it pay for the sanitation which that growth required. One can only speculate as to whether, in this tight-knit local and very hierarchical community, private representations were made from individual guardians to individual school trustees to restrict school numbers: it seems unlikely. It could be argued that that the RSA's finances were stretched to reasonable limits even before 1875. There are also two issues of comparability which should be cited in the RSA's defence. First, there is much in Chapters 2 and 3 to suggest that the Uppingham RSA was actually ahead of many of its local counterparts, notably Oakham and Stamford. Sanitary improvement was slow and patchy there too; when it came, it would certainly be expensive.[58] Second, the way in which water supply nationally was primarily provided – on a municipal basis at the start and end of the nineteenth century, but with a trend towards private operators in between[59] – suggests that we should not rush too quickly to condemn the Uppingham guardians for their resistance to the private water company in the summer of 1876. There was little consensus at that time on precisely how best to provide water to small communities, and Uppingham's RSA was not the only one to fear private monopoly providers and to face dilemmas over how far to oppose them.[60]

Underlying the RSA's problems are two further ones. Central government showed little enthusiasm for underwriting the financial problems of hard-pressed local authorities through taxation or grants rather than local rates.[61] Treasury controls – especially through high interest rates on loans – were a further obstacle, along with the PWLB's comparative slowness in responding to Uppingham's loan application. Added to this was the prevailing philosophy and practice towards RSAs of the LGB. The very tentative dealings which the Board had with the RSA both before and during the epidemic confirm its bureaucratic approach and administrative slowness and the extent to which it was wedded to the philosophy of 'local possessive pluralism',[62] even in an emergency. This arose partly out of a sense of conviction and partly because of internal tensions between its leading departments and their officials. Yet in the absence of a decision-making body at county level, the LGB was the only area or national body which could help to address local problems. The Uppingham epidemic confirms that it too was inadequate to the task it faced. With so many authorities to oversee, and such a huge range of questions from them to consider, it is scarcely surprising that it was unable to cope with rising public expectations about public health, or with being the repository for business with which other government departments did not wish to be involved. In this sense it has something in common with the modern National Health Service or government agencies responsible for social security.

The town's affairs would have been better served if Rawlinson's view that the LGB should always seek to persuade rather than coerce[63] had not been so prevalent: a feature of the affair which supports the view that the RSAs did not have sufficient enforcement powers at their disposal. The LGB ignored the repeated pleas of the Uppingham RSA for greater bye-law powers and for urban status, yet such powers might well have been a real help to it.[64] The LGB's extreme reluctance to do anything which might undermine either the RSA or Haviland is understandable – as is its feeling that this epidemic was nothing exceptional in statistical terms, and that Thring was inclined to protest too much. Yet its failure decisively to back either the school in insisting that rapid improvements were carried out, or the town in approving an immediate infusion of funds and expert back-up meant that too little was done to carry out improvements, particularly during the first three months after the school departed for Borth. It is debatable as to whether it could have killed off the disputes between Haviland and Bell if it had taken a less long-suffering, balanced and neutral view of their feud. It could have adopted a more questioning view of Haviland's report, sensing that the report would only inflame an already difficult situation. The suspicion lingers that the LGB would have dealt more effectively with Uppingham's problems if a single overall case-officer had been assigned to it (as would probably have happened in modern times). Its papers reveal that Sclater Booth, Sir John Simon, John Lambert, Robert Rawlinson and other officials all took decisions at various times, with no obviously logical pattern and no one apparently in overall control. Possibly they were too susceptible to the last visitor or correspondent seeking to lobby them. Thus the combination of an overwhelmed RSA and a reluctant and bureaucratic LGB served Uppingham's growing sanitary needs very poorly in these years.

This decade was also a highly significant time in the context of the medical issues. Knowledge about the cause of epidemic disease was tantalizingly close – and throughout the epidemic in Uppingham there seems to have been an awareness of that fact. None of the leading parties disputed that bad weather, growing populations and dirt increased the threat of disease, but none could identify its origins with precision. Nor did anyone question the idea that water supply and good drainage were the keys to safety; desire for these twin improvements can be traced right through this period. There was a general recognition of the threat from polluted water, and of the dangers of wells being too close to cesspits, but Haviland's criticisms of many of aspects of the school's administration left open the possibility of miasma or other causes.[65] If there seems to have been an absence of detailed speculation about contagion, germ or miasma theories, it is because contemporaries were inclined to believe that these processes operated in combination. In the absence of microbiological and bacteriological breakthrough, the question was: precisely how? This knowledge gap explains the references to both miasma and water-borne dangers in the findings of three of the four expert reports.[66] It would also account for Haviland's lifelong crusade against dirt in all its forms. He may well have been a 'closet miasmatist' as one contemporary reviewer suggested, although there is plenty in his many and varied criticisms of the school to suggest that he was trying to ride contagionist and miasma horses simultaneously.[67] The strident tone of his report in 1875–6, and the vehemence of his later outbursts against the siting of the waterworks near the sanatorium and the deficiencies in other parts of the town suggest a man prepared to use invective to mask a lack of definitive causal evidence, a suspicion confirmed by the judgements of Revd Talbot in the Isle of Man in the 1880s and Professor Barrett much more recently.[68]

Little survives of any detailed speculation by Dr Bell, Thring or the clergy-men on these issues – if it ever existed. This might suggest that epidemiological knowledge in the capitals of Europe was indeed slow to filter down into the pop-ular imagination in rural areas, although it also could be argued that Thring had more pressing concerns than to muse on such matters: after all, he was no expert in them. While both Thring and Bell probably deserve criticism for failing to see the potential threat posed by the early cases of typhoid, and for their failure to stop boys visiting other houses or to isolate suspected cases early enough, we must remember that guidelines for school doctors were a feature of the 1880s and subsequent decades, rather than of the 1870s.[69] Thring and Bell were prob-ably also right to fear that sending pupils home would merely risk spreading the disease across England, as well as increasing the threat to the school's reputation and prospects. There is plenty of evidence that Uppingham was not alone as a boarding school in facing the threat of epidemics, or in its inability to identify and implement ways of preventing them. Thring's protests were not against the diagnosis of how to reduce risk: only against the blame being placed entirely on the school, and against the reluctance of Barnard Smith and Haviland to accept that a boarding school could not operate a medically ideal set of arrangements – for example, through housemasters spending unlimited sums in a short space of time on their houses, or in isolating pupils there for long periods. The lim-ited extent to which the existing water supply was analysed for impurities and the reluctance of the RSA to sanction a new, private one tend to support the idea that Uppingham's epidemic came a little too early (i.e. before the 1880s) for decisive action to be taken to eliminate either bad drainage or foul water supplies.[70]

Uppingham was a town whose medical market was unusually fierce. Two decades after the regulation of qualifications had been achieved, the provision of country GPs was well developed: in this small town Dr Bell had daily to com-pete with his rival GPs for the custom both of individuals and of the school. In Dr Walford in particular, there was probably a ready potential replacement for Bell as school doctor. Drs Walford and Brown would have been only human if they wondered periodically in these years as to whether either of them might be called on to replace Dr Bell as school doctor – or whether in such circumstances Dr Childs would gain the post. The arrival of an MOH as able and messianic as Haviland posed an additional and very real threat to Bell's position – as did the recruitment by Thring of Childs as sanitary officer at Borth. Bell's regular disputes over territorial matters, both with Haviland and Childs, confirm the ferocity of the competition and the extent of the pressure on him.[71] He was certainly defensive in the face both of parents and of the RSA, and we cannot be sure how open he was with Thring himself. He appears to have been inadequate to the task he faced, both in his initial diagnoses and preventive actions, but typhoid was hard to diagnose with certainty, and the practical implications of confining pupils to houses were formidable. Sending them home to spread the infection far and wide was scarcely an inviting prospect for him. There was little accessible literature to guide him. Once the crisis developed, Bell did all he could to protect the school's position, but it can be argued that his actions were selfishly in his own interests as well as those of the school as his client/employer. His approaches to the LGB and his other arguments with Haviland – over the latter's summons to the GPs to attend meetings or over his tendency to visit their patients – were often obstructive. Yet in many ways Bell's territorial jealousies were no worse than the personality clashes between Thring and his various opponents, and they reflected the negative popular perception of the public health movement as the province of busybody inspectors who threatened the long-held notion that an Englishman's home (or school) was indeed his castle.[72] There are similarities here with the way in which attitudes have changed over time about the correct level of official involvement in other educational matters – for example, in the greatly increased public concern about child protection issues which has led to the social services inspectorate's regular dealings with boarding schools.

Haviland's actions and attitudes deepen our understanding of the role of the new rural MOHs appointed in this period. He too had a reputation to protect: even as a health official only recently arrived in the area, some of the blame would stick to him (by association in the public mind, at least) if the epidemic spiralled out of control. His passionate commitment to sanitary reform was undoubtedly genuine, yet his actions made few concessions to the logistical difficulties which the school faced in restricting the movement of pupils, and he made concerted agreement much harder to achieve. It can be argued that he was doing precisely what his job required him to do – and that the fatalities which occurred so soon after the pupils' return to school in the autumn of 1875 confirmed the school's inadequate response to the June case. But he had a huge area to police; he lived far away from Uppingham, and once he left the town after his periodic visits, he did not have to live with the consequences or fallout of his actions.

It would be hard to exaggerate the impact of the hostility between Haviland and Bell. They were quite different in temperament: Haviland, the campaigning crusader with thwarted surgical ambitions, pitted against Bell, the doughty and defensive guardian of his medical territory; Bell, the needling introvert, versus Haviland, the fiery extrovert with an eye for his audience or readership. They also represented the polarities of contemporary medical training: Haviland had become an 'academic' doctor as well as a public health official, while

Bell, content to remain a country GP, exemplifies the slowness with which new medical knowledge and ideas filtered down into the localities. However, the enmity between Bell and Haviland was only one of several rivalries running in Uppingham. It is all too easy to neglect the power of individual personalities in small, close-knit and sometimes claustrophobic communities. Uppingham had more than its share; indeed, it could be argued that the town was far too small to accommodate such an assortment of egos as those of Thring, Wales and Haviland, and, to a lesser extent, Barnard Smith and Bell. Ultimately, just as Bell was unlucky to find himself pitted against such an able and forceful MOH as Haviland, so the two rectors were singularly unfortunate to encounter such a dynamic and imaginative headmaster as Thring.

Thring, Barnard Smith and Wales had in common an ability to inspire confrontation, rather than conciliation or compromise. Barnard Smith was an experienced RSA chairman by 1875, so we have to presume that he felt he had a clear view of the speed at which rates could be allowed to rise, even in the face of the sort of crisis which the school experienced in that year. He could reasonably have felt that his bursarial career at Peterhouse had given him just as much practical experience of building matters as that which Thring had acquired over the years. He was not a school trustee, and he did not live in the town; he was poles apart from Thring temperamentally and in many other ways, but he appears to have been less directly confrontational towards the headmaster than William Wales. Wales and Thring were of very differing personalities and churchmanship. Wales's irritation with the clergymen housemasters for declining to take a greater share of services in the parish church is likely to have been indicative of a wider tension – fanned perhaps by the hardships of his own childhood, and by his interest over many years through the SPCK in the education of a class of young people untouched, even excluded, by the fashionable new boarding schools.

In large communities, ambitious and powerful leaders tend to be more dispersed; in rural areas the large number of roles undertaken by a comparatively small number of individuals increase the risk of multiple conflicts of interest. This applied particularly to the guardians, many of whom were also farmers or traders *and* ratepayers.[73] The complex relationship between the town and the school reveals a network of personal and professional rivalries which the typhoid crisis was bound to strain to the limit. Rector Wales and the RSA clerk W. H. Brown (later to be found professionally wanting) held multiple roles and exercised much power both formally and informally in the local community: the same could be said both of individual guardians and school trustees. Wales's conflict of interest as both a leading RSA figure and a school trustee would surely cause more than passing comment today. Professional competition acquired an additional dimension, for example, in the law practice of William Sheild, where two partners may have found themselves on opposite sides: Sheild, as one of the guardians who was supportive of Barnard Smith, and Pateman, who favoured the school. The epidemic also put old friendships at risk: for example, between Hodgkinson and Wales. There were conflicts of loyalty for others, too – especially those loyal to Wales as their rector, yet opposed to him as guardian. These included John Hawthorn and, later, churchwarden William Compton. Not all the shopkeepers were dissenters; a few were stalwarts of Wales's parish church. It needs little imagination to picture the professional pressures on someone such as J. C. Guy, the bank manager, who no doubt had to deal with supporters of both sides – or the personal fears of those who held land from the rector, whose influence in both real and less tangible ways extended so widely.

While no specific evidence has emerged of traders keen to replace rivals as suppliers to the school and its houses, human nature would surely have ensured that such hopes existed. Housemasters would have had the discretion to choose where to place their patronage in a town with a large number of suppliers of certain goods. Thring would certainly have had the power to influence housemasters over such choices and to put some shops out of bounds to pupils if individual shopkeepers failed to back him. Yet for townspeople the disapproval of the omni-influential Wales was not something to be risked lightly either. He could almost be said to exemplify Wohl's 'tight little oligarchy' in his own person, given his many and varied roles, although that label is perhaps more aptly applied to the guardians as a whole.[74] As the crisis developed during 1876, many of the traders and small businessmen had increasingly to balance their desire for limits on rate increases with their desire for sanitary improvement and their fear of what a prolonged absence by the school would do to their businesses.[75] The extent to which agricultural recession was already being felt is hard to quantify, and instances of individual business bankruptcies may be hard to pin down, but local fears were real enough.

Thus the Uppingham epidemic demonstrates with great clarity the importance of local circumstances, and the limitations of national government in imposing its will on localities if dominant local people were opposed to these policies, to the speed at which their implementation was proposed, or indeed to each other as individuals. The clashes of all such personalities militated against radical change and towards the perpetuation of the status quo.

What of Thring himself? Historians of Victorian education have mostly seen the Borth adventure as a pivotal event in his career: one which marked the end of a period of sustained battling both in Uppingham (to get his school built and fully established) and externally (against the Endowed Schools' Commission) before a final decade in which his achievements and reputation were beyond dispute. He kept the forces of philathleticism at bay, at a time when other schools were increasingly embracing them.[76] Some emphasize his extreme single-mindedness and determination in the face of adversity as key factors in this success.[77]

Thring's obituary article in the *Stamford Mercury* quoted the anonymous 'W', who had written to the *Pall Mall Gazette*: 'Uppingham has lost its second founder and England perhaps her ablest and certainly her most original educationalist since Arnold of Rugby ...'. After listing all his other achievements during thirty-four years in Uppingham, 'W' concluded:

> He might have been a great soldier if he had not been a great schoolmaster; for he was a born leader of men. This characteristic was never more forcibly illustrated than in 1876 – a feat, considering the magnitude of the undertaking and the risks which it involved, unprecedented in the annals of English education.[78]

There is no doubt that the events of 1875–7 show Thring's ability to use his iron determination born out of desperation, his organizational ability and energy, and his imaginative and visionary qualities to the full. Considering all that he had to do to get the school established in Borth in so short a time, the way in which he simultaneously kept up the pressure on events in Uppingham is remarkable. Even so, direct comparisons with Arnold are arguably impossible or pointless. The two men faced different challenges in different eras and they had different personal priorities. The things which they had in common were a sheer force of personality which could make both staff and pupils awestruck, and the magnetism and ability to make many of their fellow headmasters look to them as the leaders of their generation. We cannot know precisely how Arnold would have responded to a similar crisis at Rugby, or whether his staff would have responded to it in similar ways as Thring's masters did. One thing *is* now clear. By previously concentrating so much on Thring's adventure in Borth itself, historians have hitherto paid too little attention to the events in Uppingham which formed its backdrop. Thring's vituperative attacks on the RSA are understandable, given the situation he faced, and he was prepared to use outside his contacts in a way not open to the RSA, but as a result of the fierce invective of Thring and his well-connected supporters, historians have gained a one-sided view.

The traditional view of these events as the struggle of a victimized school against an incompetent and uncaring town, driven by its desire to put Thring in his place,[79] is far too simplistic. The LGB papers, the RSA minute book and Dr Bell's 'Letterbook' all need to be used as balancing agents against Thring's diaries and letters, the *USM* and the subsequent writings of his disciples. Thanks to the records preserved by Uppingham School, the Record Office for Leicester, Leicestershire and Rutland and the National Archives, and through the work of a group of very active local historians, Uppingham provides us with what may well be a uniquely detailed study of a rural community in crisis, and a reminder of the immense problems and struggles involved in bringing about the provision of the universal public utilities which many of us so often now take for granted. In this respect the scale of Thring's gamble, and of his achievement, is indeed unique – supporting his judgement, a few years later: 'That year at Borth stands alone in the history of schools'.[80]

APPENDIX 1: UPPINGHAM UNION MEMBERSHIP

N.B: The most frequent attendees are listed first, as a separate group.

Name	Place of home	Occupation	Attendance 4:75 –9:75	10:75 –3:76	4:76 –9:76	10:76 –1:77	TOTAL	Misc
Foster	Uppingham	Solicitor/ Landowner	22	19	2		43	
Parker	Preston	Farmer	10	9	10	6	35	
Rooke	Gretton	Farmer	19	7	16	11	53	Ed/SC
Sheild	Uppingham	Solicitor	6	15	13	7	41	SC
Simkin	Wardley	Farmer (died 1/76)	12	4			16	SC/VC
Smith	Glaston	Rector	26	27	22	12	87	Ch/Ed/SC
Wales	Uppingham	Rector	6	13	12	7	38	Ed/eo/SC/UT
Woodcock	High/Add St	Baker/ Greengrocer?	20	23	20	14	77	
Wortley	Ridlton/Brooke/ Oakham Rd	Farmer	4	10	17	8	39	Ch 77/SC/VC
Baines	Ridlgton/Seaton	Farmer	5	4	5	5	14	
Bell	High St	Surgeon/Dr			14	11	25	SC 76
Berry	Medbourne	Farmer	1	1			2	
Bryan	Stoke Dry	Farmer	2	1	1		4	
Burton	Drayton	Farmer	5	10	8	1	24	SC
Clarke	High St	Blacking Manu-facturer	3	2	2	2	9	
Corry	?				3		3	
Dennis	Luffenham	Clergy						eo/SC
Evans	Bisbr. Hall	Landowner	1	4	5	1	11	eo/Ed/SC/UT
Grimsdick	Slawston	Farmer			2	1	3	
Hay	Beaumont Chase	Farmer						
Henwicke	?		1				1	
Holland	Drayton	Farmer	8	7			15	
Johnson	Bisbroooke	Farmer		4	4	2	10	
Letts	Medbourne	Farmer	3	5	3		11	
Marchant	Easton Magna	Farmer		5	4	4	13	
Mould	Easton Magna	Farmer			4	2	6	
Piercy	Slawston, Lcs	Clergy				1	1	eo/SC
Pridmore	S Luffenham	Farmer	1	3	5	5	14	
Pretty	S Luffenham	Farmer	1				1	

Name	Place of home	Occupation	Attendance				TOTAL	Misc
			4:75 –9:75	10:75 –3:76	4:76 –9:76	10:76 –1:77		
Robinson	Oakham Rd	Glass/China/ Corn	3		2	2	7	
Royce	Laxton/Oakham?	Farmer	1	1	2	3	7	
Sanders	?		2				2	
Satchell	Gretton/ possibly Uppingham	Farmer/ Builder/ Caldecott Inn	1	4	4	4	13	
Sharman	?					2	2	
Shelton	Barrowden	Farmer/ Wheel Inn	6	5	3	2	16	
Simkin	Hallaton	Gentleman/ Farmer			3	1	4	OT/SC 76
Thompson	Stoke Dry	Clergy		2	1		3	eo/SC
Wade	Wardley	Farmer	4	2	6			SC 76

Source: Record Office of Leicestershire, Leicester and Rutland, DE1381/441.

Key:

Ch = Chairman	Ed = Education Committee
eo = ex officio	OT = Trustee of Oakham School
SC = Sanitary Committee	UT = Trustee of Uppingham School
VC = Vice Chairman	

APPENDIX 2: ABSTRACT OF SUMS RAISED BY RSAs

UPPINGHAM   RSA NO. 413

| | Contributions | | | | CONTS |
	General Expend	Special Expend	Party Grants MOH/ Insp Nuisance	All Other Receipts	Total
1874	402	161	1,737		2300
1875	202	276	79	34	591
1876	202	474	81	38	795
1877	408	819	67	180	1,474
1878	200	703	90	140	1,133
1879	215	759	93	57	1,124
1880	206	452		9	667
1881	218	501	161	88	968
1882	231	508	93	514	1,346
1883	215	521		233	969

| | General Expenses | | | Special Expenses | | |
	Salaries	Hospitals	Other expenses	Sewer Construction	Water Provision	Lighting
1874	174		20	1,886		
1875	237		62		83	
1876	190		49	50	35	
1877	249		124	0 plus special loan 2,075	255	
1878	295		48	63 + 890	191	
1879	285		29	62 + special loan 434	30	
1880	237		15	6 + loan 5	13	
1881	385		32	9 + special loan 98	27	
1882	263		189		40	
1883	191	37	31	166	12	

| | | | | | EXPS | |
	Other Expenses	Loans Repaid	Total	RSA Rate Value	New Loans	Loan Outstanding
1874	291		2,351	96,291		1,600
1875	151	106	639	97,015	400	1,947
1876	505	134	963	97,176		1,880
1877	392	132	3,227	99,000	3,300	4,813
1878	318	330	2,125	99,435		4,647
1879	320	329	1,489	99,938	500	4,980
1880	267	355	898	99,897		4,797
1881	270	354	1,175	104,395		4,611
1882		345+281	1,118	110,996		4,428
1883	272	388	1,097	not given		4,244

OAKHAM   RSA NO.412

	Contribution		Parly Grants MOH/ Insp Nuisance	All Other Receipts	CONTS Total
	General Expend	Special Expend			
1874	304				304
1875	202	11			213
1876	2	2	23		27
1877	204	179	47		430
1878	204	333	46	23	606
1879	204	803	46	19	1,072
1880	82	253	46	1	382
1881	104	442		23	569
1882	119	501	43	21	684
1883	23	454	31	12	520

	General Expenses			Special Expenses				Loans Repaid	Total	EXPS		
	Salaries	Hospitals	Other expenses	Sewer Construction	Water Provision	Lighting	Other Expenses			RSA Rate Value	New Loans	Loan Outstanding
1874	86		86						172	92,658		
1875	55		15	11					81	92,658		
1876	176		7		2				185	92,826		
1877	189		13	110			59		371	95,803		
1878	179		11		146		179		515	95,953		new loan 2,875
1879	228		14	10 + 1,396 special loan	110		127	250	2,135	95,982	1,000	3,746
1880	209		11	9 + loan 45	15		201	341	831	96,643		4,097
1881	97		11	14 + loan 3	9		234	291	659	96,715	500	3,948
1882	169		15	121	5		150	286	746	108,049		3,799
1883	135		7	15	6		204	281	648	not given		3,650

MARKET HARBOROUGH   RSA NO.402

| | Contributions | | | | CONTS |
	General Expend	Special Expend	Parly Grants MOH/Insp Nuisance	All Other Receipts	Total
1874	161	215		49	425
1875	165	489	86	9	749
1876	288	335	87		710
1877	177	163	82		422
1878	148	296	89	39	572
1879	247	277	89	95	708
1880	196	462			658
1881	99	369			468
1882	207	311		80	598
1883	168	251		82	501

| | General Expenses | | | | Special Expenses | | | | | | EXPS | | |
	Salaries	Hospitals	Other expenses	Sewer Construction	Water Provision	Lighting	Other Expenses	Loans Repaid	Total	RSA Rate Value	New Loans	Loan Out-standing
1874	169		38	237					444			
1875	202		154	381	33		5		775	147,173		
1876	202			411	15				628	162,745		
1877	215		58	234	7				514	163,314		
1878	232		29	183 +641					1,085	163,475		600 new loan
1879	232		36	481	1			39	789	163,915		580
1880	230		38	345 +215	25			40	893	164,996	300	860
1881	214		40	336	27				673	133,887		829
1882	200		28	312	23			125	688	137,618		734
1883	193		35	318	32			124	702	not given	300	967

MELTON MOWBRAY   RSA NO.411

	Contributons				CONTS
	General Expend	Special Expend	Partly Grants MOH/ Insp Nuisance	All Other Receipts	Total
1874	213	233		6	452
1875	213	262		19	494
1876	214	469		17	700
1877	231	53		10	294
1878	215	338		9	562
1879	222	288		18	528
1880	221	266		11	498
1881	148	115		7	270
1882	250	230			480
1883	nil	142		19	161

	General Expenses			Special Expenses						EXPS		
	Salaries	Hospitals	Other expenses	Sewer Construction	Water Provision	Lighting	Other Expenses	Loans Repaid	Total	RSA Rate Value	New Loans	Loan Outstanding
1874	120		22	133					275			
1875	190		5	299	52		24		570	136,359		
1876	270		9	65	126		12		482	136,359		
1877	260		5	76	157		2		500	137,141		
1878	60		2	394			57		513	137,604		
1879	213		3	186			53		455	137,604		
1880	270		10	174			51	24	529	137,798		303
1881	110		3	69	5			39	226	141,035		280
1882	204		11	250	32		1	38	536	160,145		257
1883	143		6	23	11		57	36	276	not given		233

STAMFORD   RSA NO.414

| | Contributions | | | | CONTS |
	General Expend	Special Expend	Party Grants MOH/Insp Nuisance	All Other Receipts	Total
1874	573			30	603
1875	193			114	307
1876	194			112	306
1877	197			110	307
1878	197			108	305
1879	197			100	297
1880	203			93	296
1881	198	37	93		328
1882	211		64	1	276
1883	219		194	112	525

| | General Expenses | | | Special Expenses | | | | | EXPS | | | |
	Salaries	Hospitals	Other expenses	Sewer Construction	Water Provision	Lighting	Other Expenses	Loans Repaid	Total	RSA Rate Value	New Loans	Loan Out-standing
1874	235		36						271	91.556		
1875	274		14						288	92,800		
1876	267		13						280	93,225		
1877	273		12						285	94,766		
1878	269		20						289	94,520		
1879	259		15						274	94,614		
1880	237		20						257	94,921		
1881	249	765	27				33		1,074	94,962	850 (a)	850
1882	303	34	30				41	51	459	101,113		836
1883	265	5	12				58	51	391	not given	440 (b)	1,261

Source: Accounts and Papers: Local Government Taxation – Abstract of Sums Raised and Expended by Rural Sanitary Authorities, and Parliamentary Local Taxation Returns, LXII–LXVII (1873–84).

General and Special expenditure refer to rate-borne costs. Some additional sums spent via loans have been included in the expenditure figures. The reasons for Stamford's low spending are explained in Chapter 3.

Key:

(a) = to provide cemetery £765
(b) = to add to cemetery £346

NOTES

Introduction

1. Similar schools for girls would be created in later decades, but it would be well into the post-1945 period before the leading boarding schools became co-educational.

2. This was the body responsible for the governance of the school. See Chapter 2.

3. See T. Traylen, *Turnpikes and Royal Mail of Rutland* (Stamford: Spiegl Press, 1982), *Oakham in Rutland* (Stamford: Spiegl Press 1982) and *Uppingham in Rutland* (Stamford: Spiegl Press 1982). See also Uppingham Local History Studies Group (ULHSG), *Canon Aldred's Historical Notes* (Uppingham, 1999), *Uppingham in 1802: A Year to Remember?* (Uppingham, 2002) and *Uppingham in 1851: A Night in the Life of a Thriving Town* (Uppingham, 2001).

4. C. Hamlin, 'Mudding in Bumbledom: On the Enormity of Large Sanitary Improvements in Four British Towns 1855–1885', *Victorian Studies*, 32 (1988), pp. 55–83. This article deals with four medium-sized urban communities: Barnsley, Birmingham, Merthyr Tydfil and Leamington Spa.

5. P. W. J. Bartrip, *Mirror of Medicine: A History of the British Medical Journal* (Oxford: Clarendon Press, 1990), pp. 52–7.

6. Hamlin, 'Mudding in Bumbledom', p. 57. This theme is echoed by Bartrip, *Mirror of Medicine*, pp. 53–4, who cites Tom Taylor's remark as secretary of the Local Government Act Office, that local government was 'the rule of unmitigated selfishness and penny wisdom under the specious mask of local liberty'. The *BMJ* was concerned that GPs themselves were sometimes forced to pay for patient improvements which it believed that the guardians should have provided.

7. A. Digby, 'The Local State', in E. J. T. Collins (ed.), *The Agrarian History of England and Wales, VII, 1850–1914*, 2 vols (Cambridge: Cambridge University Press, 2000), vol. 2, pp. 1425–64, on p. 1428; F. M. L. Thompson, 'Landowners and the Rural Community' in G. E. Mingay (ed.), *The Victorian Countryside*, 2 vols (London: Routledge, 1981), vol. 2, pp. 457–74, on pp. 458–9.

8. G. E. Mingay, *Rural Life in Victorian England* (London: Heinemann, 1977), p. 167.

9. R. J. Morris, 'The Middle Class and British Towns and Cities of the Industrial Revolution 1780 to 1870', in D. Fraser and A. Sutcliffe (eds), *The Pursuit of Urban History* (London: Edward Arnold, 1983), pp. 286–305, on p. 295. See also R. J. Morris, 'The Middle Class and the Property Cycle during the Industrial Revolution', in T. C. Smout (ed.), *The Search for Wealth and Stability: Essays in Economic and Social History Presented to M. W. Flinn* (London: Macmillan, 1979), pp. 91–113, on p. 110, and R. J. Morris,

'The Friars and Paradise: An Essay in the Building History of Oxford 1801–1861', *Oxoniensia*, 36 (1971), pp. 72–98, on p. 95.

10. For example, see Merthyr (sewerage) and Wakefield (water supply) – in Merthyr's case despite very hostile legal action from a local industrialist. See Hamlin, 'Mudding in Bumbledom', pp. 64–5. See also J. D. Marshall, *Furness and the Industrial Revolution: An Economic History of Furness (1711–1900) and the Town of Barrow (1757–1897) with an Epilogue* (Barrow-in-Furness: Barrow-in-Furness Library and Museum Committee, 1958), p. 410.

11. E. P. Hennock, 'Finance and Politics in Urban Local Government in England 1835–1900', *Historical Journal*, 6:2 (1963), pp. 212–25, on pp. 217–18.

12. R. Millward and S. Sheard, 'The Urban Fiscal Problem, 1870–1914: Government Expenditure and Finance in England and Wales', *Economic History Review*, 48:3 (1995), pp. 501–35, on pp. 527, 505.

13. Hennock, 'Finance and Politics', pp. 217–18.

14. S. Szreter, *Health and Wealth: Studies in History and Policy* (Rochester, NY: University of Rochester Press, 2005), p. 282.

15. S. Webb and B. Webb, *Statutory Authorities for Special Purposes* (London: Longmans, Green and Co., 1922), pp. 477–84. See also *ODNB*, vol. 57, pp. 810–5; and R. G. Hodgkinson, *The Origins of the National Health Service: The Medical Services of the New Poor Law 1834–1871* (London: Wellcome Historical Medical Library, 1967), p. 288, in which she cites similar concerns about local guardians.

16. See J. Garrard, *Leadership and Power in Victorian Towns 1830–1880* (Manchester: Manchester University Press, 1983). He studies three northern towns: Rochdale, Bolton and Salford.

17. See, for example, A. Hardy, *The Epidemic Streets: Infectious Disease and the Rise of Preventive Medicine 1856–1900* (Oxford: Clarendon Press, 1993); D. Watkins, 'The English Revolution in Social Medicine 1889–1911' (PhD dissertation, University of London, 1984); E. P. Hennock, *Fit and Proper Persons: Ideal and Reality in Nineteenth Century Government* (London: Edward Arnold, 1973). See also A. Wilkinson, 'The Beginnings of Disease Control in London' (DPhil dissertation, Oxford University, 1980). For details of the major epidemic in Croydon, see N. A. Cambridge, 'The Life and The Times of Dr Alfred Carpenter 1825–1892' (DMed dissertation, University of London, 2002). For civic rivalries, see K. Goschl, 'A Comparative Study of Public Health in Wakefield, Halifax and Doncaster 1865–1914' (PhD dissertation, Cambridge University, 1999).

18. A. S. Wohl, *Endangered Lives* (London: Methuen, 1983).

19. B. Reay, *Microhistories: Demography, Society and Culture in Rural England, 1800–1930* (Cambridge: Cambridge University Press, 1996), p. 72. See also A. Howkins, 'Rural Society and Community: Overview', in Collins (ed.), *The Agrarian History of England and Wales*, vol. 2, pp. 1501–14, on p. 1510; E. T. Hurren, '"The Bury-al Board": Poverty, Politics and Poor Relief in the Brixworth Union, Northamptonshire c 1870–1900' (PhD dissertation, Leicester University, 2000), and 'Poor Law versus Public Health: Diphtheria, Sanitary Reform and the "Crusade" against Outdoor Relief 1970–1900', *Social History of Medicine*, 18:3 2005), pp. 399–418. See also the developing work of the University of Leicester Centre for Urban History, twenty miles to the west of Uppingham: www.le.ac.uk/urbanhist.

20. A. Everitt (ed.), *Perspectives in English Urban History* (London: Macmillan, 1973), p. 235.

21. A. Howkins, 'Types of Rural Community', in Collins (ed.), *The Agrarian History of England and Wales*, vol. 2, pp. 1297–353, on p. 1336. See also F. M. L. Thompson, 'An Anatomy of English Agriculture, 1870–1914', in B. A. Holderness and M. Turner, *Land, Labour and Agriculture, 1700–1920* (London: Hambledon Press, 1991), pp. 211–40, on pp. 216–17.

22. T. W. Bamford, *Rise of the Public Schools: A Study of Boys' Public Boarding Schools in England and Wales from 1837 to the Present Day* (London: Nelson, 1967), p. 194.

23. Ibid., pp. 192–3. See also Bamford's study of the town and school situated in Rugby, Warwickshire: 'Public School Town in the Nineteenth Century', *British Journal of Education Studies*, 5 (1957), pp. 25–36. Bamford observed: 'In some cases, like Oundle and Uppingham, the situation is practically the same today. School depressions were town depressions – times of unrelieved gloom, with unemployment, empty houses, and lounges deserted in hotels' (p. 192).

24. S. E. Finer, *The Life and The Times of Sir Edwin Chadwick 1800–1890* (London: Methuen, 1952); S. Halliday, *The Great Stink of London: Sir Joseph Bazalgette and the Cleansing of the Victorian Capital* (Stroud: Sutton Publishing, 2001).

25. See R. Lambert, *Sir John Simon 1816–1904 and English Social Administration* (London: Macgibbon and Kee, 1963); and W. M. Frazer, *Duncan of Liverpool* (London: Hamish Hamilton, 1947).

26. K. T. Hoppen, *The Mid-Victorian Generation 1846–1886* (Oxford: Oxford University Press, 1998), pp. 91–124.

27. There is a large volume of work on the LGB, e.g. C. Bellamy, *Administering Central-Local Relations, 1871–1919: The Local Government Board in its Fiscal and Cultural Context* (Manchester: Manchester University Press, 1988) esp. pp. 111–65. See also O. MacDonagh, *Early Victorian Government 1830–1870* (London: Weidenfeld and Nicolson, 1977). The early years of the LGB are described by Lambert, *Sir John Simon*; and in R. M. Macleod, *Treasury Control and Social Administration: A Study of Establishment Growth at the Local Government Board 1871–1905*, Occasional Papers on Social Administration, 23 (London: Bell, 1968); R. M. Macleod (ed.), *Government and Expertise: Specialists, Administrators and Professionals 1860–1919* (Cambridge: Cambridge University Press, 1988). See G. Sutherland, *Studies in the Growth of Nineteenth-Century Government* (London: Routledge and Kegan Paul, 1972).

28. *Hansard*, 205 (3 April 1871), pp. 1115–43. George Joachim Goschen (1831–1907) was President of the Poor Law Board 1868–70, First Lord of the Admiralty 1871–4, later Chancellor of the Exchequer.

29. Most members of these boards were elected. This significance of this becomes clear in Chapter 7.

30. Hurren, 'Poor Law versus Public Health', pp. 403–4.

31. JPs continue to play an important local role in the UK as magistrates to this day, dealing with a variety of smaller legal offences.

32. See Chapter 2.

33. See Bellamy, *Administering Central-Local Relations*, pp. 2–3: 'Most academic comment on the [LGB] ... has taken up uncritically the views of [Sir] John Simon, the Webbs and other early twentieth century advocates of a statist model of central-local government relations, which was a counter-thesis to the more influential pluralistic nineteenth century model, without adequately noticing its dialectical context, because it happens to fit the social democratic ideology of twentieth century academic social policy ... At least until the First World War, much more political and intellectual effort was dedicated

to defining and patrolling the boundaries of the central-local divide, limiting the incursions of localities into national politics, and, especially, restricting their demands on the national Exchequer.'

34. Ibid., p. 7. See also Millward and Sheard: 'The Urban Fiscal Problem', p. 526.

35. Macleod, *Treasury Control*, pp. 24–37. See also Szreter, *Health and Wealth*, pp. 286–7.

36. Ibid., p. 273, 274.

37. Hurren, 'Poor Law versus Public Health', and '"The Bury-al Board"'.

38. *The Times*, 5 December 1878: 'In the cottage sanatorium', according to a letter from 'A'.

39. J. R. de S. Honey, *Tom Brown's Universe: The Development of the Victorian Public School* (London: Millington, 1977), p. 164.

40. P. Horn, 'Country Children', in Mingay (ed.), *The Victorian Countryside*, vol. 2, pp. 521–30, on p. 525.

41. *Medical Officer* (28 May 1938), p. 224.

42. M. McCrum, *Thomas Arnold: Head Master – A Reassessment* (Oxford: Oxford University Press, 1989), p. 37.

43. Honey: *Tom Brown's Universe*, p. 164.

44. Medical Research Council, *Epidemics in Schools: An Analysis of the Data Collected during the First Five Years of a Statistical Inquiry by the School Epidemics Committee* (London: HMSO, 1938), pp. 20–1.

45. T. H. Simms, *The Rise of a Midland Town: Rugby, 1800–1900* (Rugby: Borough of Rugby Library and Museum Committee: Rugby, 1949), pp. 5–7.

46. G. H. O. Burgess, *The Curious World of Frank Buckland* (London: Baker, 1967), p. 57.

47. Medical Research Council, *Epidemics in Schools*, p. 20. However, it contrasts Winchester's failings with the excellent attention to sanitary detail at Shrewsbury (p. 25).

48. Ibid., p. 41.

49. Honey, *Tom Brown's Universe*, p. 164, based on *Lancing Register 1848–1912*, (1913), p. xxxi. However, whereas the Uppingham outbreak was almost certainly due to foul water, Lancing's resulted from food: infected cream from a local dairy had been served at a summer cricket match against its former pupils. There are many references in the Medical Research Council report, *Epidemics in Schools*, to poor food in schools: see pp. 20–3.

50. Medical Research Council, *Epidemics in Schools*, pp. 34, 41.

51. Ibid., p. 39.

52. Honey, *Tom Brown's Universe*, p. 165. The return of diphtheria baffled the medical profession, many of whom had not previously experienced it. See also C. Creighton, *A History of Epidemics in Britain*, 2 vols (Cambridge: Cambridge University Press, 1891–4), vol. 2, pp. 723–6.

53. Creighton, *A History of Epidemics*, vol. 2, p. 723.

54. Medical Research Council, *Epidemics in Schools*, p. 18.

55. *George Moberly's Journal 1849*, ed. C. A. E. Moberly (London: John Murray, 1916), p. 75.

56. Medical Research Council, *Epidemics in Schools*, p. 25. There was no sanatorium until 1868, and all patients were nursed in the housemaster's private wing.

57. Honey, *Tom Brown's Universe*, p. 164.

58. Medical Research Council, *Epidemics in Schools*, pp. 41, 35.

59. See P. Bennett, *A Very Desolate Position* ([Fleetwood]: Rossall School, 1977).

60. See the article by Clement Dukes in *Private Schoolmaster* (15 November 1887).

61. A. L. Irvine, *Sixty Years at School* (Winchester: P. and G. Wells, 1958), pp. 20–1. The victim in one such case, a Winchester boy, had contracted the disease twice in one year at school.

62. Medical Research Council, *Epidemics in Schools*, p. 39.

63. Ibid., p. 33. The housemaster allowed pupils to go home once the epidemic broke out.

64. H. R. F. Pyatt, *Fifty Years of Fettes* (Edinburgh: Edinburgh University Press, 1931), pp. 93–100.

65. Medical Research Council, *Epidemics in Schools*, p. 41.

66. D. Newsome, *A History of Wellington College* (London: John Murray, 1959), pp. 215–2. Thirty-two boys were withdrawn. The subsequent improvements were supervised by Rogers Field: see p. 371.

67. D. Leinster-Mackay, *The Rise of the English Prep School* (London: Falmer Press, 1984), pp. 122, 123–4, 125.

68. The term 'public school' means something very different in the UK to what it denotes in, for example, the USA. British public schools are run independently of government and charge fees. Thus, confusingly, they are effectively 'private' schools.

69. Medical Research Council, *Epidemics in Schools*, p. 27.

70. Ibid., p. 27.

71. 3 March 1876, in G. R. Parkin (ed.), *Edward Thring, Headmaster of Uppingham School: Life, Diary and Letters*, 2 vols (London: Macmillan 1898), vol. 2, p. 36. For Rugby's move, see Bamford, *Rise of the Public Schools*, p. 205. It took place in 1841: some went to Churchover and Leamington instead.

72. J. B. Hope Simpson, *Rugby since Arnold: A History of Rugby School from 1842* (New York: St Martin's Press, 1967), p. 10.

73. Bamford, *Rise of the Public Schools*, p. 204. The school was hit by epidemics in 1846, 1848 and 1861.

74. Ibid., p. 208.

75. Medical Research Council, *Epidemics in Schools*, p. 42.

76. See Chapter 5.

77. C. Dukes, *Health at School considered in its Mental, Moral and Physical Aspects* (London: Rivington, 1887), p. 63. There had been a school doctor at Rugby School since 1868. Dukes held the post from 1871 until 1908.

78. G. F. Browne, *Recollections of a Bishop* (London: Smith, Elder, 1915), p. 56.

79. Honey, *Tom Brown's Universe*, p. 166.

80. Browne, *Recollections*, p. 55.

81. K. F. Kiple (ed.), *The Cambridge Historical Dictionary of Disease* (Cambridge: Cambridge University Press, 2003), pp. 345–6.

82. R. L. Huckstep, *Typhoid Fever and other Salmonella Infections* (Edinburgh: Livingstone, 1962), p. 17.

83. For references to the state of contemporary knowledge, see J. M. Eyler, *Sir Arthur Newsholme and State Medicine 1885–1935* (Cambridge, Cambridge University Press, 1997), p. 42; A. Hardy, *Health and Medicine in Britain since 1860* (Basingstoke: Palgrave, 2001), pp. 33–4; W. W. C. Topley and G. S. Wilson, *Principles of Bacteriology, Virology and Immunity* (London: Edward Arnold, 1975), p. 414; F. B. Smith, *The People's Health 1830–1910* (London: Weidenfeld and Nicolson, 1979), pp. 247–8; E. Hart, *Waterborne Typhoid: A Historic Summary of Local Outbreaks in Great Britain and Ireland 1858–1893* (London: Smith, Elder, 1897), p. 4; W. Budd, *Typhoid Fever: Its Nature, Mode of Spreading, and Prevention* (London: Longmans and Green, 1873), preface; A.

M. Anderson, *The Antiseptic Treatment of Typhoid Fever* (Dundee: John Leng, 1892), p. 5; Wohl, *Endangered Lives*, p. 89; *The Times*, 2 and 6 January 1875; *Lancet* (28 August 1875).

84. Eyler, *Sir Arthur Newsholme*, p. 42.
85. C. Hamlin, 'Politics and Germ Theories in Victorian Britain: The Metropolitan Water Commissions of 1867–9 and 1892–3', in Macleod (ed), *Government and Expertise*, pp. 111–23, on pp. 113–14.
86. Hardy, *Health and Medicine*, pp. 33–4.
87. Topley and Wilson, *Principles of Bacteriology*, p. 414.
88. Smith, *The People's Health*, pp. 247–8.
89. *BMJ* (26 February 1876).
90. *The Times*, 13 January 1876; Registrar General's Annual Reports, see Hart, *Waterborne Typhoid*, p. 4.
91. Budd, *Typhoid Fever*, preface; *BMJ* (6 May 1871).
92. Anderson, *The Antiseptic Treatment of Typhoid Fever*, p. 5.
93. Wohl, *Endangered Lives*, pp. 1–2.
94. *The Times*, 2 and 6 January 1875.
95. M. Pelling, *Cholera, Fever and English Medicine 1825–65* (Oxford: Oxford University Press, 1978), pp. 59–60.
96. Budd, *Typhoid Fever*, p. 146.
97. Wohl, *Endangered Lives*, p. 89.
98. *Lancet* (28 August 1875).
99. Dukes, *Health at School*, p. 267.
100. Ibid., pp. 264–5, ascribed to Dr Parkes.
101. A. Digby, *The Evolution of British General Practice 1850–1948* (Oxford: Oxford University Press, 1999), p. 18: 'When the whole town turned out for the funeral of a well-loved and respected GP, it was a sure sign that (he) had been a prime exponent of the art of manipulating a local environment in order to construct an ecological niche within the community'.
102. See Wilkinson, 'The Beginnings of Disease Control', pp. 23–8.
103. J. L. Brand, *Doctors and the State: The British Medical Profession and Government Action in Public Health 1870–1912* (Baltimore, MD: Johns Hopkins Press, 1965), pp. 146–9.
104. He was operating in a 'medical market'. See Digby, *The Evolution of British General Practice*, esp. pp. 93–125; and A. Digby, *Making a Medical Living: Doctors and Patients in the English Market for Medicine, 1720–1911* (Cambridge: Cambridge University Press, 1994), pp. 107–34. Chapter 3 gives further details.
105. For a general discussion of this point, see Digby, *The Evolution of British General Practice*, p. 63.
106. Pelling, *Cholera*, p. 310.
107. For details of contemporary MOHs, see C. Hamlin, 'Sanitary Policing and the Local State, 1873–1874: A Statistical Study of English and Welsh Towns', *Social History of Medicine*, 18:1 (2005), pp. 37–61, on p. 40.
108. Although his role in this is disputed: see, for example, L. Strachey, *Eminent Victorians* (London: Penguin Classics, 1986), pp. 200–2.
109. E. B. Castle, *Moral Education in Christian Times* (London: Allen and Unwin, 1958), p. 312.
110. A. Percival, *The Origins of the Headmasters' Conference* (London: John Murray, 1969), p. 21.

111. D. Newsome, *Godliness and Good Learning: Four Studies on a Victorian Ideal* (London: John Murray, 1961), p. 220. It is only fair to admit, however, that not everyone sees Thring in that light. T. W. Bamford believed that Arnold was much the greater man because, more than any of the Victorian headmasters, he strode a national stage (Bamford, *Rise of the Public Schools*, p. 162). See also B. Simon and I. Bradley, *The Victorian Public School: Studies in the Development of an Educational Institution* (Dublin: Gill and Macmillan, 1975), p. 71: 'Beside [Arnold], his nearest rivals pale into insignificance, Thring was almost a nonentity in his narrowness'.

112. A. C. Tait, successor to Arnold at Rugby, campaigned strenuously for improvements there, only (in 1856, after he had left to become Dean of Carlisle) to lose five of his own children to scarlet fever within five weeks.

113. G. Hoyland, *The Man Who Made a School: Thring of Uppingham* (London: SCM Press, 1946), p. 84 – but Hoyland's hagiographical book is far from the best of those so far written on Thring.

114. Parkin (ed.), *Edward Thring*. Bound copies of *USM* are available in UA.

115. J. H. Skrine, *Uppingham by the Sea* (London: Macmillan, 1878). For other details, see the Works Cited.

116. B. Matthews, *By God's Grace: A History of Uppingham School* (Maidstone: Whitehall Press, 1984); D. Leinster-Mackay, *The Educational World of Edward Thring* (London: Falmer Press, 1987).

117. Matthews, *By God's Grace*, pp. 108–9; Leinster-Mackay, *The Educational World of Edward Thring*, p. 12.

118. M. D. W. Tozer, 'Thring at Uppingham-by-the-Sea: The Lesson of the Borth Sermons', *History of Education Society Bulletin*, 36 (Autumn 1985), pp. 39–44; Thring to Parkin 10 April 1877, in Parkin (ed.), *Edward Thring*, vol. 2, pp. 66–8.

119. Bamford, *Rise of the Public Schools*, pp. 205–6.

120. A. Percival, *Very Superior Men: Some Early Public School Headmasters and their Achievements* (London: C. Knight, 1973), p. 191. Thring had been under pressure from critics in the years up to 1875 over his refusal to espouse the philathleticism which was becoming popular in schools comparable to Uppingham. See Chapter 8.

121. Lloyd to Hoyland, 18 April 1945, in UA. Hoyland's manuscript would become *The Man Who Made a School*.

1 Town and School

1. J. Bourne, *Understanding Leicestershire and Rutland Place Names* (Loughborough: Heart of Albion Press, 2003), p. 112.

2. W. Page (ed), *The Victoria History of the County of Rutland*, 2 vols (London: Constable, 1908–35), vol. 2, p. 247.

3. 1871 census: RG 10/3301–2.

4. P. Clark (ed.), *The Transformation of English Provincial Towns 1600–1800* (London: Hutchinson, 1984), p. 48.

5. Based on census returns. But see also R. Field, 'Report to the Sanitary Authority' (6 January 1876), in UA. On p. 2 of that report he suggests slightly lower figures.

6. ULHSG, *Uppingham in 1851*, p. 7; B. Matthews, *The Book of Rutland* (Buckingham: Barracuda Books, 1978), p. 100. The 1861 figure, which does not include the 360 or so school pupils on holiday at the time when the census was taken. By contrast, the 1871 figure (taken in term-time) included the school personnel.

7. ULHSG, *Uppingham in 1851*, p. 42; Traylen, *Turnpikes and Royal Mail of Rutland*, p. 171.
8. Page (ed.), *The Victoria History of the County of Rutland*, vol. 2, pp. 95–103.
9. A. Rogers, *The Making of Uppingham as Illustrated in its Topography and Buildings* (Uppingham: ULHSG Publication, 2003), p. 1.
10. A. Mee, *Leicestershire and Rutland* (Rotherham: King's England Press, 1997), p. 279.
11. B. Newman, *Portrait of the Shires* (London: Hale, 1968), p. 143.
12. *USM* (1911). There had once been quarries along Stockerston Road.
13. Slater's *Directory of Leicestershire and Rutland* (1850); Harrod's *Directory of Leicestershire and Rutland* (1870); Barker's *Leicestershire and Rutland Directory* (1875); Kelly's *Directory of Leicestershire and Rutland* (1876); C. N. Wright's *Commercial and General Directory of Leicestershire and Rutland* (1880), all held in Record Office of Leicestershire, Leicester and Rutland. According to the 1801 census, 94 of the 1,393 residents of Uppingham were directly and mainly engaged in farming – including graziers and smallholders, millers and maltsters. ULHSG, *Uppingham in 1802*, pp. 5–7.
14. J. Hopewell, *Shire County Guide to Leicestershire and Rutland* (Princes Risborough: Shire, 1984), p. 63.
15. Traylen, *Uppingham*, p. 3; R. Palmer, *Folklore of Leicestershire and Rutland* (Wymondham: Sycamore Press, 1985), p. 127.
16. Kelly's *Directory*.
17. *USM* (1913): the writer was a housemaster, Howard Candler.
18. Palmer, *Folklore*, p. 164.
19. Ibid., pp. 1, 3.
20. Ibid., p. 154.
21. ULHSG, *Uppingham in 1802*, p. 1.
22. Palmer, *Folklore*, p. 267.
23. ULHSG, *Uppingham in 1802*, p. 65. Baines's sweetshop had been in existence for well over a century by this time.
24. 77 of 143: see Kelly's *Directory*; Slater's *Directory*. It is likely that this continuity, coupled with the large number of small family businesses, would have created tight networks of local trust and an atmosphere of economic stability in the town: for a discussion of this issue, see S. Nenadic, 'The Small Family Firm in Victorian Britain' *Business History*, 35:4 (1993), pp. 86–114.
25. ULHSG, *Uppingham in 1851*, p. 7.
26. Ibid., p. 8.
27. See later in this chapter for details of the Sheild legal partnership.
28. There is evidence of a Norman church before the fourteenth century: information from Peter Lane.
29. Harrod's *Directory* – although the Statement of Expenses in the Restoration and Enlargement of Uppingham Church (Record Office of Leicestershire, Leicester and Rutland (ROLLR), DE5430) suggests that it was nearer £6,000.
30. ULHSG, *Uppingham in 1802*, p. 41.
31. On Wales, see H. I. Longden, *Northamptonshire and Rutland Clergy from 1500*, 16 vols in 6 (Northampton: Archer and Goodman, 1938–52), vol. 14, p. 123. See Chapters 4 and 7 for Compton's decisive interventions in the events to come.
32. J. P. Graham, *Forty Years of Uppingham* (London: Macmillan, 1932), pp. 75–6; ULHSG, *Canon Aldred's Historical Notes*. Churchwardens and sidesmen were the leading unpaid volunteers in each parish.

33. ULHSG, *Uppingham in 1851*, pp. 46–8; ULHSG, *Uppingham in 1802*, p. 18. Some, however (e.g. Compton), were supporters of both church and chapel. Methodism arrived in 1817; the dissenting community grew steadily through the century, in the years before Wales's arrival. There was no Roman Catholic Church, nor other denominations.

34. ULHSG, *Uppingham in 1802*, p. 2; D. Newton and M. Smith, *The Stamford Mercury: Three Centuries of Newspaper Publishing* (Stamford: Shaun Tyas, 1999), p. 178. The paper reverted to Toryism again in the late 1880s under a new editor.

35. ULHSG, *Uppingham in 1802*, p. 3.

36. *Stamford Mercury*, 1 January 1875.

37. *Stamford Mercury*, 26 March 1875.

38. ULHSG, *Uppingham in 1802*, p. 62.

39. *Stamford Mercury*, 23 April 1875.

40. ULHSG, *Uppingham in 1802*, p. 62.

41. Wright's *Directory*.

42. Harrod's *Directory*. It is not clear if the reading room and classrooms were ever built.

43. Slater's *Directory*. It is uncertain as to whether they were still in existence.

44. Kelly's *Directory*. Many of the magistrates were also trustees of the school.

45. ULHSG, *Uppingham in 1851*, p. 26. He had changed his family name from Gilson to Sheild, in order to inherit this property.

46. His brother, Frederick, was one of the town doctors – see Chapter 3.

47. See Chapter 2.

48. Kelly's *Directory*.

49. Traylen, *Uppingham*, p. 17; ULHSG, *Uppingham in 1802*, p. 53. Wright's *Directory*, makes reference to the Gas Company, which had as its directors Messrs Hodgkinson, Hart, Compton, Sheild and Pateman. It moved to Gas Hill in 1867.

50. ULHSG, *Uppingham in 1851*, p. 37; *USM* (1885). See also Traylen, *Uppingham*, p. 20.

51. ULHSG, *Uppingham in 1851*, p. 27.

52. Electricity appears in Matkins *Almanack* for 1924: 'Oakham Gas and Electricity Co'. It was 1932 before Canon Aldred persuaded his Parochial Church Council to equip the parish church. See ULHSG, *Canon Aldred's Historical Notes*.

53. Everitt (ed.), *Perspectives in English Urban History*, p. 235.

54. 1901 census data: S. A. Royle, 'The Development of Small Towns in Britain', in M. Daunton (ed.), *The Cambridge Urban History of Britain, Vol. III: 1840–1950* (Cambridge: Cambridge University Press, 2000), pp. 151–84, on p. 169.

55. Traylen, *Uppingham*, p. 8; ULHSG, *Uppingham in 1802*, p. 55.

56. R. C. Rome, 'Uppingham: The Story of a School 1584–1948' (undated manuscript), in UA.

57. D. Bell, *Leicestershire and Rutland Privies: A Nostalgic Trip down the Garden Path* (Newbury: Countryside Books, 2000), p. 9.

58. *USM* (March 1866); 'Mr Hodgkinson 1855–1880', *USM* (Summer 1880).

59. Rogers, *The Making of Uppingham*, p. 23.

60. Rectory Manor of Uppingham Court Rolls, vol. 6, 1782–1851, 23 October 1819, in UA.

61. ULHSG, *Uppingham in 1851*, p. 36. Ragman's Row lasted until the late 1880s, or just after.

62. According to a cutting in an Uppingham scrapbook owned by Peter Lane.

63. Field, 'Report to the Sanitary Authority', p. 2; Sir Charles Adderley to LGB, November 1875, in UA.

64. LGB notes (n.d.), in LGB Papers relating to the Uppingham Poor Law Union 1860–1882, in National Archives (NA), MH12/9815, quoting Field, 'Report to the Sanitary Authority', p. 2. Chapter 4 explains why he was reporting.

65. S. Lewis, *A Topographical Dictionary of England*, 4 vols (London: S. Lewis and Co., 1848), vol. 4, p. 420.

66. Its name is not known, although some believe that it was colloquially known as 'piss brook' at this time – possibly a play on words for the nearby village of Bisbrooke, to which it flows when it leaves Uppingham.

67. C. Jones, 'Geology in Rutland', *Rutland Natural History Society Annual Report 2000*, Rutland Library Service Local Studies ref. 508.42545 (Stamford, 2000) gives more detail.

68. Lewis, *A Topographical Dictionary of England*, vol. 4, pp. 420–1.

69. Jones, 'Geology in Rutland'.

70. *USM* (1900).

71. *Stamford Mercury*, June 1826. See also Traylen, *Uppingham*, p. 16.

72. A. Haviland, 'Report on the Geographical Distribution of Fever within the Area of the Combined Sanitary Authorities in the Counties of Northampton, Leicester, Rutland and Bucks 11 July 1874', in Northampton Central Library. This report was produced *before* the typhoid outbreak.

73. Field, 'Report to the Sanitary Authority', p. 5. It was closed in 1911.

74. B. Matthews, 'The New Water Supply', *Borth Centenary Magazine* (1977), pp. 25–6, on p. 25.

75. ULHSG, *Uppingham in 1851*, p. 48.

76. ULHSG, *Uppingham in 1802*, p. 31.

77. Rectory Manor of Uppingham Court Roll, vol. 6, 15 October 1802.

78. Manor of Preston and Uppingham Court Rolls, vol. 4, p. 2, 1863, in UA.

79. C. Rigby, 'The Life and Influence of Edward Thring' (DPhil. dissertation, Oxford University, 1968), ch. 6, p. 5.

80. Matthews, *By God's Grace*, p. 57.

81. Information drawn from *USR, 1824–1931* (sixth issue, 1932), in UA, and 1871 census.

82. *USM* (1875): he was a great friend of the school's Director of Music, Paul David. See Graham, *Forty Years*, p. 47.

83. *USM* (1885).

84. *Stamford Mercury*, 13 August 1875.

85. *USR*.

86. Graham, *Forty Years*, pp. 5, 6–7; *USM* (1885).

87. Mrs S. L. E. Haslam, Diary 1871–2, in UA.

88. B. H. Rowe, *Memoir*, quoted by Rigby in his unpublished manuscript (n.d.), ch. 14, p. 7, in UA. The words are attributed to Howard Candler.

89. *USM* (1913).

90. Bamford, *Rise of the Public Schools*, p. 195, suggests that Uppingham was not alone in this respect. The situation seems to have been much more serious in Rugby and Harrow, where 'local warfare' existed.

91. Thring, Diary 1859–62, p. 135, in UA.

92. [W. F. Rawnsley], *Early Days at Uppingham under Edward Thring by An Old Boy* (London: Macmillan, 1904), pp. 120–1.

93. Ibid., p. 119.

94. ULHSG, *Uppingham in 1851*, p. 54.

95. Ibid., p. 1. The 1851 census showed a large number of double occupations, and households headed by women. The school's growth must have created plenty of small-business opportunities.

96. Palmer, *Folklore*, p. 253.

97. *USM* (1885).

98. Graham, *Forty Years*, pp. 74–5.

99. Obituary, in *USM* (1899); letter from W. F. Rawnsley, *USM* (June 1899).

100. UA has examples: confirmed by Peter Lane.

101. Graham, *Forty Years*, pp. 75–6.

102. 1871 census. A 'boots' was a general handyman who would clean the pupils' shoes.

103. ULHSG, *Uppingham in 1851*, p. 21.

104. 1871 census. Note that the spelling is different from that of the Sheild legal family.

105. 1871 census: information about High Street West.

106. Sir Henry, first Baron Thring (1818–1907), Parliamentary Counsel, 1868–86. Thring made full use of Henry's parliamentary contacts later on – see Chapter 5.

107. For the best summary of Thring's Uppingham career, see Matthews, *By God's Grace*, chs 5 and 6.

108. See M. D. W. Tozer, 'Education for True Life: A Review of Thring's Educational Aims and Methods', *History of Education Society Bulletin* (1987), pp. 24–31, on p. 24.

109. Thring, Diary 1859–62, pp. 24–7.

110. Page (ed.), *The Victoria History of the County of Rutland*, vol. 1, pp. 261–97 gives a full discussion of how the two schools had developed.

111. Hoyland, *The Man who Made a School*, p. 36. See also Bamford, *Rise of the Public Schools*, pp. 17–38. *USR* does not list fathers by occupation, but there are references in Thring's diaries to doctor parents.

112. Rogers, *The Making of Uppingham*, p. 25; ULHSG, *Uppingham in 1851*, p. 55. These had, however, been fee-paying in recent times.

113. For discussion of this trend in similar schools, see Bamford, *Rise of the Public Schools*, pp. 198–201.

114. This is significant in view of the events described in Chapter 4.

115. In that year, in reply to a question from the secretary of the Public Schools Commissioners, he stated that these cost about £23,535 per annum. Rigby, unpublished manuscript, ch. 9, p. 3, suggests that staff costs put a very severe pressure on Thring financially.

116. Matthews, *By God's Grace*, p. 85.

117. *USR*, 1875.

118. Graham, *Forty Years*, pp. 53–73. *USM* portraits usually appeared when a member of staff retired or died. For example, George Mullins at West Deyne, a keen meteorologist and bee-keeper, and William Vale Bagshawe, a lover of music and keen fisherman. Theophilus Rowe lectured to school and town alike on topics as varied as the surface of the moon, life in a lighthouse and marks on snail-shells.

119. Hoyland, *The Man Who Made a School*, p. 46.

120. Honey, *Tom Brown's Universe*, pp. 297–8: 'At Uppingham Thring, who had no margin in the endowment to expand his school, guaranteed his masters to supply them with boys if they would build (for perhaps £8,000) boarding houses as private profit-making ventures'.

121. For example William Earle, Usher (Second Master) under Holden, and brother of Walter Earle who built Brooklands on the London Road.

122. D. Tate, 'West Deyne: A Short History' (undated manuscript), in UA; Matthews, *By God's Grace*, p. 109. Hodgkinson moved to run the Lower School in 1867. He built a house in the grand style of a French chateau, at a cost of no less than £12,000.

123. Ibid., p. 84.

124. Thring, Diary 1859–62, p. 36.

125. Wright's *Directory*.

126. Hodgkinson at the Lower School was also reliant on Thring's recommendation to parents.

127. Wright's *Directory*. Thring himself was paid £200 per year plus a capitation fee for each boy by 1880.

128. For example, J. R. Blakiston: see Rigby, 'The Life and Influence of Edward Thring', ch. 6, p. 130.

129. See J. Venn and J. A. Venn (comps), *Alumni Cantabrigienses: A Biographical List of all Known Students, Graduates and Holders of Office at the University of Cambridge, from the Earliest Times to 1900* (Cambridge: Cambridge University Press, 1922); J. Foster, *Alumni Oxonienses* (Oxford: Parker and Co., 1888). Matthews, *By God's Grace*, p. 180: Sam Haslam ran his house for 37 years, George Mullins and George Christian for 33 years and William Campbell for 32.

130. W. F. Rawnsley, *Edward Thring: Maker of Uppingham School, Headmaster 1853–1887* (London: Kegan Paul, Trench, Trubner and Co., 1926), p. 22.

131. Parkin (ed.), *Edward Thring*, vol. 2, pp. 201–2. See also Page (ed.), *The Victoria History of the County of Rutland*, vol. 1, p. 296.

132. Parliamentary Papers: Annual Reports of the Charity Commissioners: 1875, XX 13, and 1876, XX 19. The figure was unchanged over several years, suggesting that the funds may not have been very actively managed.

133. Parkin (ed.), *Edward Thring*, vol. 2, pp. 189–93. See also Page (ed.), *The Victoria History of the County of Rutland*, vol. 1, p. 294.

134. Rigby, unpublished manuscript, ch. 11, p. 6. One master, W. F. Witts (who had left Uppingham by 1875), contributed £1,000. The masters were also required by the trustees to contribute at least £3,000 towards the new schoolroom. The choice of architect was one of many issues on which Thring clashed with the trustees.

135. For details, see Matthews, *By God's Grace*, pp. 93–4, 98.

136. *USM* (1900): Hodgkinson and Candler started donations to a fund to buy property which could be knocked down to make way for new fives courts in the early 1870s. Thring, Mullins and Haslam followed their example, and over £850 was raised in all. Candler gave selflessly – including £1,220 used to buy a games field on the Leicester Road which he then leased back to the school. Meanwhile another £3,000 had been spent on a sanatorium.

137. Matthews. *By God's Grace*, pp. 94, 106.

138. Kelly's *Directory*.

139. H. D. Rawnsley, *Edward Thring: Teacher and Poet* (London: T. Fisher Unwin, 1899), pp. 91–2. At the time of the Tercentenary Appeal in 1884, Thring stated that £91,000 had been expended over thirty years.

140. This description of Thring was provided by Cormac Rigby in conversation, 2002. C. W. Cobb, a housemaster, wrote (many years later) that 'matters of ecclesiastical ceremony, of procedure or of posture were to him unimportant'. *USM* (1926).

141. For details of this event, see Percival, *The Origins of the Headmasters' Conference*.

142. Welldon of Tonbridge to Mitchinson of the King's School, Canterbury, in Parkin (ed.), *Edward Thring*, vol. 1, p. 199.

2 Local Society and Local Government

1. For further details, see Page (ed.), *The Victoria History of the County of Rutland*, vol. 2, p. 247; also Palmer, *Folklore*, p. 210; Kelly's *Directory*. Rutland was occupied by 13,000 cattle, 110,000 sheep and about 6,000 pigs – a level of livestock which caused poachers to be a problem, especially of deer, pheasants or hare. Besides land for grazing there was also some forest and a few local quarries, as well as areas of nursery and market garden. Crops included wheat, barley, oats, rye, peas and beans – as well as potatoes, turnips and swedes, carrots and cabbages. Stilton cheese was also produced locally.

2. E. Walford, *The County Families of the United Kingdom* (London: R. Hardwicke, 1876), pp. 111–18.

3. Westmorland, at 1:449,000, was at the other extreme.

4. J. V. Beckett, *The English Aristocracy 1660–1914*, (Oxford: Blackwell, 1986), pp. 52–3.

5. J. Bateman, *The Great Landowners of Great Britain and Ireland 1876* (London: Harrison and Sons, 1878) p. 178. Lord Gainsborough had 15,076 acres in Rutland, as well as 3,500 acres in 5 other counties. Fludyer, Finch and Wingfeld had, respectively, 2,638/2,000+, 9,183/8,200 and 2,905/600+, spread across 5 other counties.

6. Bateman observed that Rutland more than most counties was not good at 'holding on to its acres' (i.e. ensuring that these great landowners concentrated most of their landholdings within the county). Despite the magnificent hunting character of the locality, he identified 16 significant owners of Rutland land whose main estates lay in other counties, and whose total land holdings amounted to an area equivalent to nearly 40 per cent of the county itself. They included the Duke of Rutland (whose vast estates were mainly in Leicestershire and Derbyshire), the Marquis of Exeter (just across the Lincolnshire border at Burghley House, Stamford) and three other school trustees. Two of them lived locally: Edward Conant at Lyndon Hall, with 1,500 acres in Rutland plus nearly 4,000 acres in Lincolnshire; and George Watson at Rockingham Castle, in Northamptonshire but within sight of Rutland across the Welland valley. The third was the chairman of the trustees, A. C. Johnson, whose lands were in Lincolnshire at Wytham-on-the-Hill. All these may be assumed to have had a strong vested interest in restricting local tax levels. See Bateman, *Great Landowners of Britain and Ireland*, p. 101.

7. The 1851 census listed 21 smaller farmers in Uppingham, whose average holding was just over 56 acres. Only 2 of them had more than 6 employees. There were also several dozen allotment holders: ULHSG, *Uppingham in 1851*, p. 18.

8. Land tax assessment 1874–5, ROLLR, supplied by Peter Lane.

9. Charles George Noel, the second Earl (1818–81).

10. *The Times*, 15 August 1881, and information from Lady Sarah Campden: 'Though he represented Rutland in the Liberal interest, he is ranked in *Dod* of that year as a Conservative'. The Whigs were one of the two main political parties in England at the time. Opposed to the Tories/Conservatives, they would later evolve into the Liberal party.

11. *The Times*, 15 August 1881.

12. ULHSG, *Uppingham in 1802*, p. 73, and information from Lady Sarah Campden.

13. His uncle was also named C. B. Adderley.

14. *The Times*, 29 March 1905.

15. Bateman, *Great Landowners of Britain and Ireland*, p. 527, where he is described as an 'owner-occupier'. Bateman mentions the difficulties in separating out personal and institutional holdings for clergymen, and states that glebe lands are entered under the names of the incumbent. The chancellor of a diocese dealt with a wide variety of church legal matters on behalf of the bishop.

16. Return of Owners of Land, vol. 2: Rutland (HMSO, 1873), ROLLR, supplied by Peter Lane.

17. Information from Peter Lane, based on Rectory Manor of Uppingham Court Rolls, vol. 6, and Episcopal Visitation: Uppingham, 1878, Northamptonshire Record Office, ML598. These were considerable. The 1875 visitation shows separate accounts for home, foreign and local purposes of nearly £20 each. Copyhold was a type of land tenure, the title deeds being a 'copy' of the record of the manor court.

18. ULHSG *Uppingham in 1802*, p. 74; ULHSG, *Uppingham in 1851*, p. 26.

19. ULHSG, *Uppingham in 1851*, pp. 26, 52.

20. Return of Owners of Land, vol. 2.

21. Bell to W. T. Jacob, 1 September 1876, in T. Bell, 'Letterbook' (1876–1904), in UA.

22. ULHSG, *Uppingham in 1851*, p. 23: see the example of the decline of William Hopkins and family.

23. See Rutland Local History and Record Society, 'Who was Who in Rutland', *Rutland Record Society*, 8 (1988), in Rutland County Museum, Oakham; A. Jenkins, *Rutland: A Portrait in Old Picture Postcards* (Seaford: S. B. Publications, 1993). Finch was an archetypal landed man: a keen huntsman who entered parliament as Tory MP for Rutland in 1867, held the seat for 40 years and defeated the Liberal candidate in the 1906 re-election with the slogan 'the agricultural candidate for an agricultural constituency'. Hon. W. C. Evans-Freke at nearby Bisbrooke Hall did not own quite enough to make Bateman's lists, but he was both a trustee and a local guardian. Mr Edward Dawson could have been included in Bateman's list of Rutland, but for the fact that Launde Abbey lay just across the Leicestershire border.

24. M. Stenton (ed.), *Who's Who of British Members of Parliament*, 4 vols (Hassocks, Harverster Press, 1976–81), vol. 1, 1832–85. He was Conservative MP for Rutland, 1847–83, Lord of the Treasury, 1866–8, Parliamentary Secretary to the Treasury, 1868, and Chief Commissioner of Public Buildings and Works, 1876–80. Noel was officially a Whig, although with distinctly conservative leanings according to Lady Sarah Campden.

25. In addition to the rector, the church was represented by the Bishop of Peterborough (Rt Revd William Conor Magee), and the Dean (Very Revd Augustus Saunders). Oxford and Cambridge universities nominated one trustee each.

26. Parkin (ed.), *Edward Thring*, vol. 1, p. 192.

27. See also Wright's *Directory*, p. 529.

28. Matthews, *By God's Grace*, p. 87.

29. Kelly's *Directory*: Adderley, Conant, Evans-Freke, Finch, Fludyer, Noel, Wales, Watson and Wingfield.

30. Uppingham School Trustees' Minute Books, in UA.

31. G. Carnell, *The Bishops of Peterborough 1541–1991* (Much Wenlock: R. J. L. Smith, 1993), p. 80: 'Creighton's Primary Charge (as bishop) showed his awareness of contemporary economic pressures and their human consequences'.

32. Rawnsley, *Edward Thring*, p. 20. According to L. von Glehn (Creighton), *Life and Letters of Mandell Creighton – by his Wife*, 2 vols (London: Longmans, Green, 1904), vol. 2, p. 24, Creighton spoke at a church meeting in Leicester in 1891: 'The towns,

as their streets grow, tell of the activity of the municipal body, of the care of sanitary inspectors. But it is left to the quickened conscience of the community at large to do what is needed to maintain the high spiritual interests, without which eternal things are vain and empty ....'

33. Thring, Diary 1859–62, p. 95.

34. S. Fletcher, *Feminists and Bureaucrats: A Study in the Development of Girls' Education in the Nineteenth Century* (Cambridge: Cambridge University Press, 1980), p. 7. See also F. E. Balls, 'The Origins of the Endowed Schools Act 1869' (PhD dissertation, Cambridge University, 1964), p. 448.

35. Matthews, *By God's Grace*, pp. 86–7: in his early years Thring clashed particularly with Sir Gilbert Heathcote, Baronet, squire of Normanton in Rutland, whom he described variously as a man of 'sordid character ... [who] showed his usual narrow bigotry' in what he said at meetings; 'a great conceited baby'; 'a running sore in the body', and 'anything more ignorant I've never heard'.

36. Rigby, unpublished manuscript, ch. 11, p. 1.

37. Matthews, *By God's Grace*, p. 87.

38. See Chapter 1.

39. 1871 census, RG10/3301–2: The Rectory was at No. 2, London Road, where Wales lived in some style with his wife and sister-in-law together with 6 servants including a footman.

40. His first wife died suddenly in 1855 aged 40. His later married the Hon. Miss Spencer of Great Houghton.

41. Wales's great-uncle had worked at the school some years earlier. Wales was withdrawn in 1817, for reasons unknown, and eventually went up to St Catharine's Hall, Cambridge in 1823. I am indebted to his family descendant, Wendy Wales, for this information.

42. Longden, *Northamptonshire and Rutland Clergy*, vol. 14, p. 123.

43. Uppingham National School Minute Book, 1876, ROLLR, DE1784/64. Wales and his wife both gave generously to it despite having no children of their own. Five of Thring's housemasters also contributed generously, having had children educated there.

44. Episcopal Visitation: Uppingham, 1878, Northamptonshire Record Office, ML598. In 1875 the Visitation recorded up to 400 at Matins, 250 at the afternoon service and up to 550 at Evensong.

45. Uppingham Parish Church Restoration Fund, ROLLR, DE1784/23; ULHSG, *Uppingham in 1851*, p. 45. The 1851 religious census, a decade before Wales's arrival, had showed the strength of non-conformity: 907 adults and 156 children are said to have attended one of the 4 chapels in the town on the Sunday concerned. His successors were unable to sustain this success: see P. Lane, 'The Parish Church in Peace and at War: 1925–46', in ULHSG, *Uppingham at War: Uppingham in Living Memory: Snapshots of Uppingham in the Twentieth Century* (Uppingham, 2005), p. 41.

46. Episcopal Visitation, 1878, with a 20 per cent increase in children attending the Sunday school – although he deplored the fact that it was hard to retain them as they got older: 'So many go out (into domestic) service, both boys and girls', with the result that he believed that the bible classes were 'not very successful'.

47. Local tradition has it that he was 'Gulielmus', who worked very hard to undermine the rival chapel Sunday school in this period, and who circulated a letter containing the words: 'Everybody must know, (and sorry I am to see it) that there are more and more persons daily dissenting from the church, and I conceive it a duty incumbent upon everyone to endeavour to put a stop to their career'. See A. Peach, *A Brief Account of the*

Uppingham Congregational Church and of the Fifty Years' Ministry of Revd. John Green (Bournemouth, 1914).

48. *Northampton District Chronicle,* 21 August 1889; W. Wales, *A Lecture Delivered in the National School Room Northampton on Monday 6th May 1839, to the Members and Friends of the SPCK, Explaining the Nature and Objects of that Institution* (Northampton, 1839), in Cambridge University Library. Wales was a keen supporter of liberal education: 'Let it increasingly be felt that for a people to be uneducated is not only discreditable but disadvantageous'. He was chairman of a public meeting in Uppingham in autumn 1875 to discuss public concern about the impact of newly-arrived crowds of navvies in the area, working on the nearby new railway between Manton and Kettering. He enlisted Bishop Magee's support for a mission to them, and preached regularly at its services in the years which followed. See J. A. Paul, *3000 Strangers: Navvy Life on the Kettering to Manton Railway* (Kettering: Nostalgia Collection, 2003), pp. 72, 91–4; D. W. Barrett, *Life and Work among the Navvies* (London: Wells Gardner, Darton and Co., 1880), p. 90.

49. See Bamford, *Rise of the Public Schools,* pp. 198–201, for a discussion of this general trend.

50. Rigby, unpublished manuscript, ch. 11, p. 3. As one drives in along the A47 road from the east, the church spire and the chapel pinnacle are still very evident on the skyline.

51. There he had set up a parochial school and shown a commitment to adult education, despite some territorial jealousies from the local National Society and the largely nonconformist Mechanics' Institute. He was a visionary and strong organizer: his Religious and Useful Knowledge Society soon had a library of 2,500 volumes for newspapers and periodicals, and a small museum, and provided reading, writing and drawing classes as well as some lectures. See *Northampton Herald,* 2 April 1859: he worked in Northampton, 1832–59. See also R. J. Serjeantson, *A History of the Church of All Saints, Northampton* (Northampton: W. Mark, 1901); J. Lawes, 'Voluntary Schools and Basic Education in Northampton 1800–1871', *Northamptonshire Past and Present,* 6:2 (1979–80), pp. 85–91, describes his educational work.

52. *Northampton Herald,* 24 August 1879. The town had more than its share of free-thinkers and political radicals throughout the century, sometimes opposed to the controlling social forces of the clergy: see E. Royle, 'Charles Bradlaugh, Free Thought and Northampton', *Northamptonshire Past and Present,* 6:3 (1979–80), pp. 141–50; M. Dickie, 'Liberals, Radicals and Socialists in Northamptonshire before the Great War', *Northamptonshire Past and Present,* 7:1 (1983–4), pp. 51–4. Wales led a vigorous campaign to enforce pew rents and to collect the church rates too: see T. Milner, 'Letter to the Revd. W. Wales in Reply to that Gentleman's Sermon, by Thomas Milner M.A., Northampton, 1838', in Cambridge University Library.

53. *Northampton Herald,* 7 October 1916.

54. Evident in the civilities with which they conducted their disagreement early in 1876 – see Chapter 5.

55. ULHSG, *Uppingham in 1802,* p. 49; ULHSG, *Uppingham in 1851,* pp. 46–8. There had been strong rivalry between the congregational/independent chapel and the parish church during much of the century. He pointed out that other schools had made this commitment in similar circumstances. See also Vestry Minute Book 1869, ROLLR, DE1784/24. However, some felt that Wales was too inclined to ask others to take these services for him – including one sardonic visitor to the school who remarked: 'Wales, like England, expects every man to do *his* duty': Rigby, unpublished manuscript, ch. 11, p. 3.

56. W. Wales, 'The Minister's Duty towards Himself and His People: a Sermon preached at St Giles's Church, Northampton, at the Visitation of the Lord Bishop of Peterborough Wed. 6 August 1851', in Cambridge University Library. This address is typical of his earlier ministry: 'Remembering then that the one object of all our labours is the glory of God and the salvation of souls, let us warn, reprove, rebuke, exhort, let us plainly and fully make known the message, the mode and blessedness of that salvation which is in and by Jesus Christ'.

57. His secondary role as canon and chancellor of the Peterborough diocese (since 1850) suggests a tidy, legalistic mindset – a man who believed that forms and procedures in religion were important.

58. Point made by Peter Lane in correspondence, November 2005.

59. Uppingham Union Minute Book, ROLLR, DE1381/441.

60. Of these, 5 were clergymen, 2 derived at least part of their income from inns, 1 was a surgeon and at least 4 ran local shops or small businesses. The occupations of 4 have proved elusive. Barely half a dozen of them lived in the town itself, and few of them had any other close links with the school to keep them in touch with its affairs.

61. Uppingham Union Minute Book. There were 87 union meetings between April 1875 and early 1877. In addition to the chairman, a group of 8 predominated: Charles Simkin, union vice-chairman and a gentleman farmer from Wardley; John Woodcock, farmer and railway agent of High Street East; Samuel Rooke, a farmer from Gretton (over the border in Northamptonshire); George Foster, who lived on the Oakham Road and who farmed other lands at Brooke; William Shield, who (in addition to his manorial steward, legal and money-lending roles) was Superintendent Registrar for Births, Marriages and Deaths as well as being the local coroner; Edward Wortley of Ridlington and John Parker of Preston (farmers). The eighth was the rector, Revd William Wales. The fact that he is usually listed immediately after the chairman and vice-chairman suggests that his local status made him influential. Simkin, Foster, Rooke, Sheild, Woodcock and Wortley (from 1876) were also on its sanitary sub-committee responsible for public health matters.

62. See Introduction.

63. For Poor Law administrative purposes, England was divided up into areas known as unions. See Digby, 'The Local State', p. 1437.

64. Kelly's *Directory*: in Rutland: Ayston, Barrowden, Belton, Bisbrooke, Caldecott, Glaston, Lyddington, North Luffenham, South Luffenham, Morcott, Pilton, Preston, Ridlington, Seaton, Stoke Dry, Thorpe-by-Water, Uppingham, Wardley, Wing; in Leicestershire: Blaston St Giles/ Blaston St Michael, Bringhurst, Drayton, Easton Magna, Hallaton, Holt/Nevill Holt, Horninghold, Medbourne, Slawston, Stockerston. In Northamptonshire: Fineshade, Gretton, Harringworth, Laxton, Rockingham, Wakerley.

65. Wright's *Directory*.

66. Kelly's *Directory*.

67. ULHSG, *Uppingham in 1851*, p. 32. It had an imposing tower, a fine board room and apartments for the masters. See also Wright's *Directory*.

68. Uppingham Union Minute Book. The Registrar General's 1874 report focused on small-pox vaccinations throughout the country. Of 365 children born in the Uppingham area, 327 had been successfully vaccinated, and 31 had died unvaccinated, leaving only 7 unaccounted for – good compared with many of the other areas listed, and a feat repeated in both the following years: Registrar General's Annual Report, 1874.

69. See Appendix 2.

70. Abstract of sums raised and expended by Rural Sanitary Authorities, Parliamentary Papers 1873–4: 1875 LXII, and subsequent years. Directly comparable figures for earlier years are not available, owing to local government reorganization following the 1872 Act.

71. 1871 census information, and accounts and papers: Local government taxation – abstract of sums raised and expended by Rural Sanitary Authorities, and Parliamentary Local Taxation Returns 1874–82. See Appendix 2 for details.

72. Out-relief issues were monitored with particular care by the LGB in its early years, to see whether its directives were being implemented uniformly. See Hurren, "'The Bury-al Board'", p. 29.

73. See Chapters 4 and 7.

74. Rowe and Haslam, both of whom had boarding houses in High Street West.

75. Thring to Rt Hon. G Sclater Booth, President of the LGB, 5 November 1875, in Parkin (ed.), *Edward Thring*, vol. 2, p. 17.

76. Uppingham Union Minute Book, 29 March 1876.

77. LGB Papers, in NA MH12/9815–17, 1875–82.

78. 1871 census. See P. Pattenden, and A. Thomson, 'The Snuffing of Sanitary Smith: Fellow and Senior Bursar', *Peterhouse Annual Record 2002–3* (2005), pp. 43–56, on p. 51. Barnard Smith's activities included alms financed by tunnel money from land sold to the Midland Railway, a clothing club and a coal club. He became a member of the Rutland Deanery board, and was involved in the Rutland Society of Industry, which sought to keep families 'off the parish' through teaching children to knit and sew. He also showed an interest in public hygiene by raising money for drainage improvements to the main street (drawing no doubt on his Peterhouse experience of building contracts). He devised mathematical competitions and gave prizes for the local national and church schools. See A. Thomson, 'A Study of Roles and Relationships in a Rutland Village in the Mid Victorian Period: Glaston c1860–90' (MA dissertation, Leicester University, 1999), pp. 49–51.

79. He had been admitted to Peterhouse, Cambridge (where his family had long connections), in 1835 and was a scholar in 1838; a year later he was 28th wrangler, and a fellowship soon followed. By 1842 he was a priest, serving as dean of the college (1842–4) and as perpetual curate of Little St Mary's, before becoming college praelector and pro-proctor of the university (1860). He was the college's statutory classical lecturer from 1847 to 1855, but also successively junior and senior bursar (1840–1 and 1846–61 respectively) for an unusually long period. His period of office coincided with the development both of the buildings and of the gardens of the college, and in selling land to (amongst others) the new Eastern Counties railway. During his time the college built St Peter's terrace, re-roofed two sides of Old Court, installed new windows in the chapel, and laid a gas supply to the kitchen and to other parts of the college. It also exchanged land with the university and with Pembroke College. But he also had to deal in financial minutiae and day-to-day routine issues – including the renegotiation of payments for knife-cleaning. He had accepted his plum living from the grateful fellows on marrying at a comparatively late age in 1861. See Pattenden and Thomson, 'The Snuffing of Sanitary Smith'. His placing was amongst those with first class honours in the mathematical tripos – not fourth, as stated in Longden, *Northamptonshire and Rutland Clergy*, vol. 14, p. 207.

80. He also produced *Barnard Smith's Chart of the Metric System* (1871) and *Easy Lessons in Arithmetic* (1872). See Thomson, 'A Study of Roles and Relationships in a Rutland Village', p. 53.

81. For example, B. Smith, *Arithmetic and Algebra in their Principles and Application* (Cambridge: Macmillan, 1853), which has no introduction, and only a minimal commentary. He also wrote a pamphlet on the importance of correct procedures in matters of ecclesiastical dilapidations – a subject in which Wales, with his diocesan legal role as chancellor, would surely have taken an interest. See B. Smith, *Observations on the Ecclesiastical Dilapidations Act 1871 with a View to its Amendment or more Efficient Operation* (Cambridge, 1872). For details, see Thomson, 'A Study of Roles and Relationships in a Rutland Village', p. 47.

82. This becomes relevant when one considers whether the criticisms of the school discussed in Chapter 5 were realistic – e.g. for its failure to restrict the movement of boarders once the epidemic broke out.

83. He gave £100 to the Peterhouse hall restoration fund in 1867, and restored Glaston church as well as building a fine nine-bedroomed rectory there – another building which would attract sizeable rates. The Rectory restoration alone cost £2,000. He also contributed to the restoration of Uppingham Church, in support of his ally, Wales. *Stamford Mercury*, 6 November 1863. See also Pattenden and Thomson, 'The Snuffing of Sanitary Smith', p. 48.

84. M. J. Daunton, *House and Home in the Victorian City: Working-Class Housing 1850–1914* (London: Edward Arnold, 1983), pp. 203–6; Beckett, *The Aristocracy in England*, pp. 199–200.

85. Hoppen, *The Mid-Victorian Generation*, p. 10.

86. Railway freight loaded increased from 88m tons in 1860 to 166m in 1870, and to 232m in 1880. Ibid., pp. 289–92.

87. A. Howkins, *Reshaping Rural England: A Social History 1850–1925* (London: Routledge, 1991), esp. pp. 13, 138–9.

88. Howkins, 'Rural Society and Community', pp. 1504–5. Although it varied between regions, it resulted in 'not just falling prices but also falling incomes, declining standards of living and declining welfare for sixty-five years on end'. For a fuller discussion, see Thompson, 'An Anatomy of English Agriculture'.

89. Hennock, 'Finance and Politics', p. 216. See also Howkins, 'Rural Society and Community', p. 1504, where one writer is quoted who believes that 'Part of the depression in arable husbandry was psychological. Unease became foreboding because the traditional relationship between prices and output, between farmers and consumers, had broken down' (B. A. Holderness, 'Agricultural Responses to the "Depression" of the late Nineteenth Century in England and France: Towards a Comparative History', paper presented to IV Congresso di Storia dei Movimenti Contadini – L'Agricultura en Europa e la masita della 'questione agrarian' 1880–1914 (Rome, 1993).

90. 1865 28 & 29 Vict., c. 79. See D. Englander, *Poverty and Poor Law Reform in Nineteenth Century Britain 1834–1914: From Chadwick to Booth* (London: Longman, 1998), p. 21; Hurren, '"The Bury-al Board"', pp. 26–7; Digby, 'The Local State', p. 1438.

91. S. Webb and B. Webb, *English Poor Law History Part II: The Last Hundred Years*, 2 vols (London: Longman, Green and Co., 1929), vol. 1, pp. 429–30. See also Digby, 'The Local State', p. 1438.

92. Hoppen, *The Mid-Victorian Generation*, p. 11: imported wheat constituted half of domestic consumption by 1873–5.

93. Carnell, *The Bishops of Peterborough*, p. 80. This brought pastoral difficulties for clergy, as well as economic ones for farm workers and others.

94. Hoppen, *The Mid-Victorian Generation*, p. 22.

95. Howkins, *Reshaping Rural England*, pp. 153–5, 157.

96. Thompson, 'An Anatomy of English Agriculture', p. 212.

97. E. H. Hunt, *Regional Wage Variations in Britain, 1850–1914* (Oxford: Clarendon Press, 1973), p. 59; Thompson, 'An Anatomy of English Agriculture', pp. 222, 230.

98. Howkins, *Reshaping Rural England*, p. 13; J. Saville, *Rural Depopulation in England and Wales 1851–1951* (London: Routledge and Kegan Paul, 1957), pp. 73–4. The other counties to experience depopulation were Huntingdon and Cornwall.

99. M. Smith, *The Story of Stamford* (Stamford: Martin Smith, 1994), p. 99.

100. ULHSG, *Uppingham in 1802*, p. 23; A. Rogers, 'Prosperous – But Precarious: Mortgages in the Local Economy of Uppingham in the Eighteenth and Nineteenth Centuries', *Family and Community History*, 8:2 (November 2005), pp. 105–22.

101. Ibid., p. 116. There are suspicions of some sharp practice; solicitors arranged some mortgages to increase their local influence and hold over individuals.

102. Hardy, *Health and Medicine*, p. 12. Farr noted the link between population density and high death rates. See also W. Luckin, 'The Final Catastrophe – Cholera in London, 1866', *Medical History*, 21 (1977), pp. 32–42, on p. 42.

103. Hamlin, 'Muddling in Bumbledom'. He lists this as a catalyst.

104. Hennock: 'Finance and Politics', pp. 214, 223.

105. Goschl, 'A Comparative Study of Public Health', pp. 66, 82.

106. R. Millward and R. Ward, 'From Private to Public Ownership of Gas Undertakings in England and Wales, 1851–1947: Chronology, Incidence and Causes', *Business History*, 35:3 (1993), pp. 1–21, on pp. 1–4.

107. Lambert, *Sir John Simon*, p. 510.

108. Hoppen, *The Mid-Victorian Generation*, pp. 108–9.

109. Ibid., p. 109.

110. K. B. Smellie, *A History of Local Government* (London: Allen and Unwin, 1946), p. 30.

111. Rogers, *The Making of Uppingham*, p. 25. Uppingham's orbit stretched from Ridlington in the north to Rockingham in the south and from Duddington in the east to Slawston in the west.

112. Macdonagh, *Early Victorian Government*, pp. 129–30.

113. R. Lambert, 'The Local Government Act Office', *Victorian Studies*, 6:1 (1962), pp. 121–50, on pp. 129–35. For a full discussion, see R. K. F. J. Young, 'Sanitary Administration under the Local Government Board 1871–78' (BLitt. dissertation, Oxford University, 1964). See also Smellie, *A History of Local Government*, p. 5, for example Merthyr Tydfil in 1869.

114. Sir John Simon observed, with frustration: 'In all country districts there is one authority for every privy and another authority for every pigsty, but I do also apprehend that with regard to the privy, one authority is expected to prevent its being a nuisance, and the other to require it to be put to rights if it is a nuisance'. Young, 'Sanitary Administration', pp. 6–7. It bears close resemblance to Uppingham: see Chapter 1.

115. Bellamy, *Administering Central-Local Relations*, pp. 33–4. See also Wohl, *Endangered Lives*, pp. 171–3, for details of the 'economy' parties which sprang up in a number of towns and regions.

116. Early examples of such grants included financing of prosecutions and maintaining of prisoners, a 50 per cent subsidy for the salaries of workhouse personnel, drugs and medical equipment, and a grant to pay half the cost of salaries for the new MOHs.
117. P. J. Waller, *Town, City and Nation: England 1870–1914* (Oxford: Clarendon Press, 1983).
118. Hennock, 'Finance and Politics', p. 212.
119. Hardy, *The Epidemic Streets*, p. 168.
120. By the end of the decade, there would also be regulation of markets, street lighting and burials, and it would be compulsory to record infectious diseases.
121. Hennock, 'Finance and Politics', p. 214.
122. Smellie, *A History of Local Government*, p. 62.
123. Hurren, 'Poor Law versus Public Health', pp. 403–4; Hamlin, 'Muddling in Bumbledom', p. 67.
124. Hurren, 'Poor Law versus Public Health', pp. 401–2.
125. B. Kerr, 'Country Professions', in Mingay (ed.), *The Victorian Countryside*, vol. 1, pp. 288–99, on pp. 294, 288.
126. E.g. events in Croydon in 1852–3: see Cambridge, 'The Life and Times of Dr Alfred Carpenter', pp. 57–8. See also A. Digby, 'The Rural Poor', in Mingay (ed.), *The Victorian Countryside* vol. 2, pp. 591–602, on p. 600.
127. Wohl, *Endangered Lives*, pp. 112–16. Over the period 1848–72 local authorities had borrowed only £11m from central government for sanitary purposes: by contrast, between 1872 and 1880 loans were sanctioned totalling over £22m – a figure which rose to £66m between 1880 and 1897. An LGB survey showed that local authorities had contracted for over £3.5m worth of work by 1874, of which £2.5m was for sanitation and water supply.
128. Local Government Reports/General Papers 1875–7, in NA MH19/88, 15 April 1876. Inspectors tended to work regionally, and to be involved in 5 to 6 inspections at any one time. Major Tulloch, who came to Uppingham in July 1876, was one such. See Chapter 7.
129. See NA MH19/88, for the list for January/February 1876, and also R. C. K. Ensor, *England 1870–1914* (Oxford, Clarendon Press, 1936), p. 442.
130. J. Redlich and F. Hirst, *The History of Local Government in England* (London: Macmillan, 1958), p. 110. The process was haphazard; only with the Local Government Act of 1888 (which set up county councils) would the government have provided a sufficiently sophisticated local structure capable of coping with all these new demands.
131. Macdonagh, *Early Victorian Government*, p. 130. See also Bellamy, *Administering Central-Local Relations*, pp. 111–65.
132. Hurren, '"The Bury-al Board"', p. 28.
133. Correspondence and Papers of the LGB 1875–7, in NA MH25/26–8, give a good picture. A year later the powers of the Board of Trade and of the Home Secretary under the Highways and Turnpikes Act were transferred to the LGB too.
134. Lambert, 'The Local Government Act Office'.
135. M. R. Maltbie, 'The English Local Government Board', *Political Science Quarterly*, 13:2 (1903), pp. 232–58, on p. 236.
136. NA MH25/27, for example its ruling on 21 April 1876 that guardians should not be required to make separate provision for Jewish paupers.
137. Simon mastered vast amounts of administrative detail and his influence could be felt everywhere: '[his] small, neat handwriting is amazingly ubiquitous, remorselessly omni-

present' – as evidenced on the papers relating to Uppingham. See Lambert, *Sir John Simon*, pp. 533, 524–6.

138. Bellamy, *Administering Central-Local Relations*, p. 15. There appear to be parallels with the modern National Health Service.

139. W. M. Frazer, *A History of Public Health 1834–1939* (London: Bailliere, Tindall and Cox, 1950), p. 118. See also Szreter, *Health and Wealth*, pp. 286–7; Sir J. Simon, *English Sanitary Institutions* (London: Smith, Elder, 1897), pp. 364–5.

140. This was a genuine belief in 'mixed' (i.e. central and local) government. See Bellamy, *Administering Central-Local Relations*, p. 12.

141. Ibid., p. 116.

142. Ibid., p. 145.

143. Ibid., p. 117, 'First Report: Evidence of Robert Rawlinson'. See also C. Hamlin, *Public Health and Social Justice in the Age of Chadwick: Britain 1800–1854* (Cambridge: Cambridge University Press, 1998), p. 332; Young, 'Sanitary Administration', pp. 211–3. Rawlinson had long experience of the local authorities, including their tendency to cut costs on major projects. This sometimes had disastrous results: see G. M. Binnie, *Early Victorian Water Engineers* (London: Telford, 1981), pp. 202–22, for his involvement in the problems of the Swansea Corporation Waterworks.

144. W. A. Ross, 'The Local Government Board and After: Retrospect', *Public Administration*, 13 (1956), pp. 17–25, on p. 18.

145. Macleod, *Treasury Control*, p. 15.

146. George Sclater Booth, PC LLD FRS, Lord Basing. President of the LGB, 1874, later chairman of the first County Council of Hampshire, 1888. He was described as 'an honest and capable man of business' in his obituary; 'He had more sense than genius ... his parliamentary life may be said to have been useful rather than ambitious, solid rather than distinguished'. *The Times*, 23 October 1894.

147. G. Haw, 'The Local Government Board', *Contemporary Review*, 94 (1908), pp. 54–63, on p. 55. In the last year of the old Board (1870), 58,000 letters were received, but by 1874 the various RSAs sent not less than 84,000 to its successor, and by 1895 the figure had risen to 160,000. See also Macleod, *Treasury Control*, pp. 11, 33.

148. Bellamy, *Administering Central-Local Relations*, p. 139.

149. Ibid., p. 14.

150. Macleod, *Treasury Control*, p. 8. See also Bellamy, *Administering Central-Local Relations*, pp. 111–12.

151. Macleod, *Treasury Control*, p. 52.

152. Szreter, *Health and Wealth*, pp. 284–7.

153. Hurren, 'Poor Law versus Public Health', p. 399.

154. Sutherland, *Studies in the Growth of Nineteenth-Century Government*, p. 216.

155. M. Daunton, *Trusting Leviathan: The Politics of Taxation in Britain 1799–1914* (Cambridge: Cambridge University Press, 2001), p. 278; Bellamy, *Administering Central-Local Relations*, pp. 82, 85.

156. Public Works Loan Board Papers 1835–92, in NA MH19/190: 2nd Annual Report of the PWLB, 20 June 1877: still mostly at an interest rate of 3.5 per cent payable over thirty years – although as a result of Treasury pressure, 4 per cent or higher was charged in some cases. Much larger loans were made to major cities such as Manchester.

157. See Chapter 7.

158. Bellamy, *Administering Central-Local Relations*, p. 14.

159. R.S. Wright and H. Hobhouse, *An Outline of Local Government and Local Taxation in England and Wales* (London: Sweet and Maxwell, 1884), intro., p. viii.

3 Local Medicine and Local Doctors

1. Wohl, *Endangered Lives*, p. 91: Annual Report of the Medical Officer to the Privy Council) for 1864, appendix IV, p. 509: 'Sanitary State of Seacroft, by Dr Stevens'.
2. F. F. Cartwright, *A Social History of Medicine* (London: Longman, 1977), p. 93.
3. Hardy, *The Epidemic Streets*, p. 161.
4. See Chapter 1, and ULHSG, *Uppingham in 1802*, p. 31.
5. Ibid., pp. 91–2.
6. Daunton, *House and Home*, pp. 244–5, and Annual Report of the MOH for Nottingham (1897), p. 39.
7. Halliday *The Great Stink of London*, p. 42.
8. H. Barty-King, *Water: The Book: An Illustrated History of Water Supply and Wastewater in the United Kingdom* (London: Quiller, 1992), p. 107.
9. Wohl, *Endangered Lives*, p. 110.
10. See Chapter 1.
11. J. Priestley, 'Conservancy versus Carriage Systems for the Disposal of Excreta', *Public Health*, 7 (1894–5), pp. 280–5, on p. 280.
12. N. Williams and G. Mooney, 'Infant Mortality in an "Age of Great Cities": London and the English Provincial Cities Compared', *Continuity and Change*, 9:2 (1994), pp. 185–212, on p. 191.
13. Ibid., pp. 199, 207.
14. Ibid., p. 196.
15. J. Simmons, *Life in Victorian Leicester* (Leicester: Leicester Museums, 1971), pp. 12–14; Daunton, *House and Home*, p. 254.
16. Traylen, *Oakham*, p. 17.
17. Mr E. S. Stephens, CE, in *Stamford Mercury*, June 1868. See also Traylen, *Oakham*, p. 18, and Chapter 4.
18. Traylen, *Oakham*, pp. 19, 97.
19. J. Hassan, *A History of Water in Modern England and Wales* (Manchester: Manchester University Press, 1999), pp. 11, 25.
20. Ibid., p. 11; R. Reynolds, *Cleanliness and Godliness: or The Further Metamorphosis* (London: George Allen and Unwin, 1943), p. 99.
21. Hassan, *A History of Water*, p. 19.
22. Hardy, *The Epidemic Streets*, pp. 169–72.
23. Hassan, *A History of Water*, p. 11.
24. Ibid., p. 41.
25. J. Hassan, 'The Growth and Impact of the British Water Industry in the Nineteenth Century', *Economic History Review*, 38 (1985), pp. 521–47, on p. 532.
26. F. W. Robins, *The Story of Water Supply* (Oxford: Oxford University Press, 1946), p. 195.
27. Hassan, 'The Growth and Impact of the British Water Industry', p. 533. See also W. M. Stern. 'Water Supply in Britain: The Development of a Public Service', *Royal Sanitary Institute Journal*, 74 (1954), pp. 999–1005, on p. 999.
28. Hassan, 'The Growth and Impact of the British Water Industry', p. 543.
29. Ibid., pp. 537–8.

30. Wohl, *Endangered Lives*, p. 111.
31. Hassan *A History of Water*, p. 51.
32. Daunton, *House and Home*, pp. 246–7.
33. Cartwright, *A Social History of Medicine*, p. 95.
34. M. Falkus, 'The Development of Municipal Trading in the Nineteenth Century', *Business History*, 19:2 (1977), pp. 139–46.
35. Hurren, 'Poor Law versus Public Health', p. 408: Brixworth, Northamptonshire, in the late 1880s is a good example.
36. Hassan, 'The Growth and Impact of the British Water Industry', pp. 535, 539.
37. G. M. Howe, *Man, Environment and Disease in Britain: A Medical Geography through the Ages* (Newton Abbott: David and Charles, 1972), p. 58.
38. C. Hamlin, *A Science of Impurity: Water Analysis in Nineteenth Century Britain* (Bristol: Hilger, 1990), p. 299.
39. Hamlin, 'Politics and Germ Theories in Victorian Britain', p. 111.
40. Simmons, *Life in Victorian Leicester*, pp. 12–14.
41. A. Rogers, *The Book of Stamford* (Buckingham: Barracuda Books, 1983), p. 97.
42. *Stamford Mercury*, July 1868. See also Traylen, *Oakham*, p. 18.
43. Ibid., p. 97.
44. Digby, *Making a Medical Living*, pp. 70–7.
45. Hardy, *The Epidemic Streets*, p. 192.
46. I. Loudon, *Medical Care and the General Practitioner* (Oxford: Clarendon Press, 1986), p. 61.
47. Brand, *Doctors and the State*, pp. 146–9. See also Digby, *The Evolution of British General Practice*, pp. 48–52.
48. Ibid., pp. 63, 13.
49. Registrar General's Annual Report, 1875. See also Digby, *The Evolution of British General Practice*, pp. 190–1.
50. J. Gathorne-Hardy, *Doctors: The Lives and Work of GPs* (London: Corgi, 1984), p. 5.
51. M. R. Mitford, 'Our Village (1848)', quoted in Kerr, 'Country Professions', p. 291.
52. Digby, *The Evolution of British General Practice*, p. 31.
53. Digby, *Making a Medical Living*, p. 140.
54. Digby, *The Evolution of British General Practice*, p. 32; W. Rivington, *The Medical Profession: Being the Essay to which was Awarded the First Carmichael Prize of £200 by the Royal College of the Surgeons of England, 1879* (Dublin: Fannin and Co., 1879), p. 3.
55. Horn, P., 'Country Children', p. 525.
56. Digby, *The Evolution of British General Practice*, pp. 13, 65.
57. Ibid., p. 148.
58. Digby, *Making a Medical Living*, p. 123.
59. It was financed at least in part by a mortgage, according to an undated note in UA.
60. ULHSG, *Uppingham in 1802*, pp. 12, 81. The Leicester Royal Infirmary had opened in 1771.
61. Ibid., pp. 11–12.
62. *Medical Register* (1871). A fourth, John Bell (cousin of Dr Thomas Bell) is listed as a surgeon, but he does not appear to have been practising in the town at this time. He lived in the High Street.
63. Rivington, *The Medical Profession*, p. 2.
64. Kelly's *Directory*.
65. *USM* (June 1915).

66. ULHSG, *Uppingham in 1802*, p. 11.
67. 1871 census, RG10/3301–2.
68. *Medical Register* (1861) and subsequent volumes, and *USM* (1914).
69. *USM* (Summer 1914).
70. J. Wesley, *On Dress: A Sermon on I Peter III, 3,4* (Boston, MA: Samuel Avery, 1811).
71. *Servants' Magazine*, 2 (1839), pp. 57–9.
72. Bartrip, *Mirror of Medicine*, pp. 55–6.
73. Registrar General's Quarterly Bulletin, October–December 1875.
74. *BMJ* (26 February 1876).
75. *BMJ* (5 February 1876).
76. S. Szreter, *Fertility, Class and Gender in Britain 1860–1940* (Cambridge: Cambridge University Press, 1996), pp. 107, 190.
77. Frazer, *A History of English Public Health*, pp. 114–17.
78. Wohl, *Endangered Lives*, p. 181.
79. Hennock, *Fit and Proper Persons*, p. 114.
80. Ibid., p. 214.
81. Hodgkinson, *The Origins of the National Health Service*, pp. 649, 678.
82. Lambert, *Sir John Simon*, pp. 508–11.
83. Wohl, *Endangered Lives*, p. 181.
84. Young, 'Sanitary Administration', pp. 48–9.
85. Wilkinson, 'The Beginnings of Disease Control', pp. 13, 321. See also Watkins, 'The English Revolution in Social Medicine', p. 35.
86. Hodgkinson, *The Origins of the National Health Service*, p. 105. In 1846 Rutland had only 8 MOs, and they were poorly paid compared with those in most other counties: their total salary was only £308.
87. *ODNB*, vol. 50, pp. 660–3. Duncan was paid £750 once his post became full time.
88. Wohl, *Endangered Lives*, pp. 186–7.
89. Ibid., p. 182.
90. *BMJ* (30 May 1874).
91. North-West Association of MOHs to Sclater Booth, 29 March 1876, in Correspondence and Papers of the Local Government Board 1875–7, NA MH25/27.
92. D. Porter, *Health, Civilisation and the State* (London: Routledge, 1999), p. 136. See also Wilkinson, 'The Beginnings of Disease Control', p. 65.
93. Ibid., pp. 23–8.
94. Young, 'Sanitary Administration', pp. 174–5.
95. Wohl, *Endangered Lives*, p. 171.
96. Frazer, *A History of English Public Health*, p. 122.
97. Szreter, *Fertility, Class and Gender*, pp. 190, 197.
98. Hamlin, 'Sanitary Policing and the Local State', p. 40.
99. Venn and Venn (comps), *Alumni Cantabrigienses*.
100. Obituary article, in *BMJ* (27 June 1903); P. J. Squibbs, *A Bridgwater Diary 1800–1967* (Bridgwater: Somerset County Library, 1968), pp. 64, 106.
101. S. G. Jarman, *A History of Bridgwater* (London: Elliott Stock, 1889), p. 139.
102. Hunter's *Directory*, 1848, in Record Office of Leicestershire, Leicester and Rutland; S. G. Jarman, *The Bridgwater Infirmary: A Record of its Rise and Progress* (St Ives: Jarman and Gregory, 1890), p. 35. See also *Bridgwater Mercury*, 20 February 1867.

103. F. A. Barrett, 'Alfred Haviland's Nineteenth-Century Map Analysis of the Geographical Distribution of Disease in England and Wales', *Social Science and Medicine,* 46:6 (1998), pp. 767–81.

104. *Lancet* (28 June 1879).

105. C. Brown, *Northampton 1835–1895: Shire Town, New Town* (Chichester: Phillimore, 1990), p. 94.

106. J. C. Frost, *Haviland Genealogy* (New York: Higginson and Co., 1914), p. 14.

107. The Aftermath and Conclusion will show another example of this, some years after the Uppingham epidemic.

108. Anon., 'Review of Geographical Distribution of Heart Disease, Cancer and Phthisis by Alfred Haviland', *Athenaeum* (4 March 1876), p. 333.

109. T. W. Freeman, 'Nineteenth Century Medical Geographer', *Geographical Magazine,* 10 (1978), p. 90; F. A. Barrett, *Disease and Geography: The History of an Idea* (Toronto: York University Press, 2000), p. 325.

110. Barrett, 'Alfred Haviland's Nineteenth-Century Map', p. 767; Barrett, *Disease and Geography,* p. 323.

111. *Bridgwater Mercury,* 16 April 1873.

112. LGB Papers, in NA MH12/9815, 3 May 1873. Young, 'Sanitary Administration', p. 158, confirms that salaries ranged between £200 and £800 for most posts.

113. *BMJ* (26 April 1873), p. 480.

114. *Northampton Herald,* 13 June 1903.

115. *BMJ* (26 June 1873).

116. Young, 'Sanitary Administration', p. 137, and *BMJ* (26 June 1873).

117. Haviland, 'Report on the Geographical Distribution of Fever'.

118. For example in criticizing stream courses unsatisfactorily diverted for the new railway at Brackley, unregulated muck-heaps in Market Harborough and a burst pigsty in Oundle.

119. A. Haviland, 'Abstract of Two Lectures on The Geographical Distribution of Typhoid Fever in England and Wales: A Lecture Delivered at St Thomas's Hospital, London' *BMJ* (10 February 1872), pp. 148–9.

120. Haviland, 'Report on the Geographical Distribution of Fever', p. 7.

121. He cited the Registrar General's statistics on causes of deaths, 1875.

122. 12 in Northamptonshire, 1 in Rutland, 1 in Leicestershire, and 1 in Buckinghamshire.

123. Haviland, 'Report on the Geographical Distribution of Fever', p. 9, table 2, citing Registrar General's Annual Reports for these years.

124. A. Haviland, 'Report on the Late Outbreak of Enteric Fever in Archdeacon Johnson's School, Uppingham, Rutland: June–November 1875' (1876), pp. 35–7, in UA and CUL.

125. Ibid., p. 37.

126. Burial Registers, ROLLR, DE4862/1: only a partial indicator of deaths as a whole, as they include only those burials carried out by the Church of England.

127. *Stamford Mercury,* 9 April 1875; Registrar General's Annual Report, 1875.

128. Reports of the MOH of the Privy Council and Local Government Board, 1873.

129. MOH's Annual Report, 1874, p. 7.

130. See Chapter 4.

4 Typhoid

1. J. C. Jennings, 'Analysis of Uppingham Church Registers' (n.d.), in UA.
2. ULHSG, *Uppingham in 1851*, p. 7.
3. Traylen, *Uppingham*, p. 20.
4. Ibid., p. 20.
5. See Chapter 3.
6. Also a trustee of local charities and generous benefactor, according to Peter Lane. See Chapter 8.
7. LGB Papers, in NA MH12/9815, May 1857; *Stamford Mercury*, March 1858. See also Traylen, *Uppingham*, p. 20.
8. See Cambridge, 'The Life and Times of Dr Alfred Carpenter'. There are parallels here with Croydon in the 1850s. The Rutland Museum acquired technical drawings of these sewers and subsequent additions during the summer of 2005.
9. *Stamford Mercury*, July 1865. See also Traylen, *Uppingham*, p. 21.
10. Later Headmaster of Tonbridge. See Chapter 1 for his comments on Uppingham's water supply.
11. Thring to LGB, February 1876, in NA MH12/9815. Foster would be a guardian by 1875. The identity of the enquirer is unknown; presumably from London.
12. R. Rawlinson, 'Uppingham: Town and School Reports' (1876), p. 9, in UA.
13. J. H. Pidcock, 'A Report Presented to the Uppingham Sewer Authority Committee, on the Drainage, Water Supply etc, of the Town, by J. H. Pidcock Esq., Engineer' (May 1871), in UA.
14. Notice from Revd W Wales, Chairman of the Sewer Authority, May 1871, in NA MH12/9815.
15. Thring to LGB, February 1876, in ibid.
16. Wales request to LGB, 22 August 1871, in ibid.
17. Submission to LGB, 22 August 1871, in ibid.
18. Barnard Smith to LGB, 16 March 1872, in ibid.
19. Traylen, *Uppingham*, p. 23, based on *Stamford Mercury* reports.
20. Submission to LGB, 4 March 1872, in NA MH12/9814.
21. Whitaker and Perrott to W. H. Brown, 26 September 1872, in ibid.
22. Special meeting of the (new) Uppingham RSA, 18 September 1872, in ibid.
23. Rawlinson, 'Uppingham', p. 8.
24. This may have put him at odds with Sheild, who was a guardian. It shows how an epidemic could potentially polarize attitudes right across the local governing class.
25. John L. Pateman to LGB, 2 May 1872, in NA MH12/9814.
26. Barnard Smith to LGB, 16 March 1872, in ibid.
27. Field, 'Report to the Sanitary Authority', p. 2.
28. Rawlinson, 'Uppingham', p. 7.
29. W. H. Brown to LGB, 14 November 1872, NA MH12/9814.
30. LGB to W. H. Brown, 21 January 1873, in ibid. Poor Law retrenchment policy would have been an additional factor militating against higher spending. See also Bellamy, *Administering Central-Local Relations*, pp. 29–41, 79–94.
31. W. H. Brown to LGB, 10 March 1873, in NA MH12/9814.
32. Wales to LGB, 21 August 1872, in ibid.

33. Wright and Hobhouse, *An Outline of Local Government*, pp. 29–31. This rate would have been at 25 per cent of the land's rateable value, to allow for the fact that agricultural land would derive less direct benefit from town improvements.
34. LGB internal memorandum, 24 October 1873, in NA MH12/9815.
35. White to LGB, 8 June 1873, in ibid.
36. LGB to W. H. Brown, 24 May 1875, in ibid.
37. Brown to LGB, 1 January 1874, in ibid.
38. Inspector's report to LGB, 10 April 1874, and LGB to Brown, 20 March 1874, in ibid.
39. Brown to LGB, 14 March 1874, in ibid. The Board agreed to a 3-mile maximum.
40. Brown to LGB, 18 December 1874, in ibid.
41. RSA to LGB, 22 May 1874, in ibid. For example, West Ham in the previous year.
42. Brown to LGB, 6 July 1874, in ibid.
43. LGB to Brown, November 1874, in ibid. Precise date not known, but the LGB announced its decision to Brown on 11 December of that year.
44. Brown to LGB, 16 August 1875, in ibid.
45. Haviland to LGB, 21 October 1875, in ibid.
46. Haviland's notes, in UA.
47. Article on rainfall at Uppingham for the ten years 1874–83, in *USM* (1884).
48. Hart, *Waterborne Typhoid*, p. 4.
49. *BMJ* (5 February 1876).
50. Ibid., taken from the Registrar General's Quarterly Bulletins.
51. *BMJ* (6 May 1876).
52. Single unsigned printed sheet document of rainfall statistics, in UA.
53. Parkin (ed.), *Edward Thring*, vol. 2, p. 3. Thring was right to be worried; diphtheria was frequently fatal amongst children. See Introduction.
54. Medical Research Council, *Epidemics in Schools*, pp. 31–2.
55. Petition to LGB, 9 February 1875, in NA MH12/9815.
56. Thring's diary, 13 February 1875, in Parkin (ed.), *Edward Thring*, vol. 2, p. 4. See also Robert Hayes, Registrar's memorandum to LGB, 13 February 1875, in NA MH12/9815.
57. Thring's diary, 28 February 1875, in Parkin (ed.), *Edward Thring*, vol. 2, p. 4.
58. *Stamford Mercury*, 12 February 1875.
59. J. H. Thudicum (1829–1901), Director of the Chemical and Pathological Laboratory, St Thomas's Hospital.
60. Letter to Haviland, 12 July 1875, in UA.
61. Thring's diary, 28 February 1875, in Parkin (ed.), *Edward Thring*, vol. 2, p. 4.
62. Bell to Dr Paley, 29 January 1876, in Bell, 'Letterbook'. Typhoid diagnosis could be very difficult. See Hardy, *The Epidemic Streets*, pp. 152–3.
63. Matthews, *By God's Grace*, p. 109.
64. This was fourteen years before the Infectious Diseases Notification Act of 1889: see Chapter 3.
65. Haviland, 'Report on the Late Outbreak of Enteric Fever', p. 3, confirmed by R. J. Hodgkinson, 'Remarks on Mr Haviland's Report on the Visitation of Fever in the School and Town of Uppingham' (10 February 1876), p. 6, in UA.
66. Hodgkinson, 'Remarks on Mr Haviland's Report', p. 1.
67. Haviland, 'Report on the Late Outbreak of Enteric Fever', pp. 3, 43.
68. Ibid., p. 17.
69. *Lancet* (28 August 1875).
70. Thring's diary, 4 October 1875, in Parkin (ed.), *Edward Thring*, vol. 2, p. 4.

71. Haviland, 'Report on the Late Outbreak of Enteric Fever', p. 4.
72. List of cases, 15 September–7 November 1875, in UA.
73. Thring's diary, 6–10 October 1875, in Parkin (ed.), *Edward Thring*, vol. 2, p. 5.
74. *USM* (Autumn 1875).
75. Ibid.
76. Thring's diary, 8 October 1875, in Parkin (ed.), *Edward Thring*, vol. 2, p. 5.
77. Haviland, 'Report on the Late Outbreak of Enteric Fever', p. 28.
78. Hodgkinson, 'Remarks on Mr Haviland's Report', pp. 2–3.
79. Haviland, 'Report on the Late Outbreak of Enteric Fever', p. 28.
80. Thring's diary, 10 October 1876, in Parkin (ed.), *Edward Thring*, vol. 2, p. 5. 'The rest from work and the chapel is always a blessing'.
81. Ibid., p. 4.
82. Wright's *Directory*. See also 1871 census, RG10/3301–2.
83. Haviland, 'Report on the Late Outbreak of Enteric Fever', p. 4.
84. Thring's diary, 10 October 1875, in Parkin (ed.), *Edward Thring*, vol. 2, p. 5.
85. Thring's diary, 14 October 1875, in ibid.
86. Thring's diary, 13 October 1875, in ibid.
87. Rigby, unpublished manuscript, ch. 20, p. 3.
88. Letter to Haviland, *c.* 7 October 1875, in Parkin (ed.), *Edward Thring*, vol. 2, p. 7.
89. Thring's diary, 12 October 1875, in ibid.
90. Thring's diary, 14 October 1875, in ibid. Chapter 1 showed that dispersed boarding houses was one of the school's distinctive features.
91. Haviland, 'Report on the Late Outbreak of Enteric Fever', p. 6.
92. Thring's diary, 13 October 1875, in Parkin (ed.), *Edward Thring*, vol. 2, p. 8.
93. Haviland, 'Report on the Late Outbreak of Enteric Fever', p. 6.
94. Thring's diary, 14 October 1875, in Parkin (ed.), *Edward Thring*, vol. 2, p. 9.
95. E. Thring, 'Borth Commemoration Sermon' (1880), in UA.
96. Matthews, *By God's Grace*, p. 101. Matthews states that the total was 46.
97. List in UA.
98. Thring's diary, 18 October 1875, in Parkin (ed.), *Edward Thring*, vol. 2, p. 11.
99. Haviland, 'Report on the Late Outbreak of Enteric Fever', p. 8.
100. *The Times*, 5 November 1875.
101. Thring's diary, 16 October 1875, in Parkin (ed.), *Edward Thring*, vol. 2, p. 10.
102. Thring's diary, 22 October 1875, in ibid.
103. Letter in UA.
104. Thring's diary, 23 October 1875, in Parkin (ed.), *Edward Thring*, vol. 2, p. 12.
105. Thring's diary, 27 October 1875, in ibid.
106. Thring to Sir Henry Thring, 5 November 1875, in ibid., p. 18.
107. Letter from Prof. Frankland, early November 1875, confirmed by letter to Christian, 29 November 1875, both in UA. A decade later, with bacteriology an emergent branch of medical knowledge, the analysis might have been more extensive; see Introduction.
108. *The Times*, 5 November 1875.
109. Uppingham School Trustees' Minute Books, 29 October 1875, in UA.
110. Thring's diary, 29 October 1875, in Parkin (ed.), *Edward Thring*, vol. 2, p. 15.
111. Notice to parents, October 1875, in UA.
112. Thring's diary, 1 November 1875, in Parkin (ed.), *Edward Thring*, vol. 2, p. 15.
113. *Lancet* (6 November 1875).
114. Thring's diary, 3 November 1875, in Parkin (ed.), *Edward Thring*, vol. 2, p. 16.

5 Winter 1875–6

1. Sir Henry, first Baron Thring (1818–1907), Parliamentary Counsel, 1868–86.
2. Thring to Sir Henry Thring, 5 November 1875, in Parkin (ed.), *Edward Thring*, vol. 2, pp. 18–19.
3. Another letter went to one of his younger brothers (Revd Godfrey Thring): 'I am neither in a funk, nor cast down. I shall stand by my guns, and if knocked over, will begin again. I don't mean by begin again, try headmastership. The workhouse is open before that.' Thring to Revd Godfrey Thring, 8 November 1875, in ibid., p. 20.
4. Thring's diary, 9 November 1875, in ibid., p. 20.
5. See Chapter 2.
6. Adderley to LGB (from his London house in Eaton Place), 15 November 1875, in NA MH12/9815.
7. Thring's diary, 16 November 1875, in Parkin (ed.), *Edward Thring*, vol. 2, p. 21.
8. Tarbotton to Christian, 6 December 1875, in UA.
9. Thring to Christian, 4 December 1875, written from an address in Hyde Park, London, in UA.
10. Thring's diary, 11 November 1875, in Parkin (ed.), *Edward Thring*, vol. 2, p. 21.
11. Note of RSA meeting, 15 December 1875, in UA.
12. B. Smith, 'The Late Visitation of Typhoid Fever in the School and Town of Uppingham. A Statement of the Action of the Sanitary Committee by the Chairman 19 January 1876', p. 9, in UA.
13. Note from Hodding and Beevor, 14 December 1875, in UA.
14. Haviland to LGB, 21 October 1875, in NA MH12/9815.
15. Birley and Jacob to Thring, copied to LGB, 1 November 1875, in ibid.
16. Thring to Sir John Simon, 2 November 1875, in ibid.
17. LGB internal memorandum, 3 November 1875, in NA MH12/9816.
18. See Chapter 4.
19. Field to LGB, 3 November 1875, NA MH12/9816.
20. LGB memorandum, 3 November 1875, in ibid.
21. LGB memorandum, 3 November 1875, in ibid.
22. Thring's diary, 3 November 1875, in Parkin (ed.), *Edward Thring*, vol. 2, p. 16.
23. Note in LGB papers, 21 November 1875, in NA MH12/9816.
24. There are the initials 'GB' and the date, 7 November.
25. Memorial from Thring to LGB, 5 November 1875, in ibid.
26. LGB to RSA, 12 November 1875, in ibid.
27. Rawlinson to Lambert, Secretary of LGB, 11 November 1875, in ibid.
28. LGB memorandum, 12 November 1875, in ibid.
29. LGB papers, dates between 12 November and 17 December 1875, in ibid.
30. Bell to Sclater Booth, 12 November 1875, in ibid.
31. Bell to LGB, 7 December 1875, in ibid. The patients included a woman in advanced pregnancy.
32. LGB internal memorandum, 14 December 1875, in ibid.
33. Thring had previously had some dealings with newspapers over the years as the school's fame and reputation grew. Their coverage had not always been favourable – notably at Easter in 1861 when he had warned the boys that too many were returning late after exeat (half-term) breaks, citing difficulty in finding suitable trains as an excuse. After he had caned two boys who had ignored this warning, the father of one of them wrote

demanding an apology and threatening to complain to the trustees. Other parents in the boy's home area started to take sides; the father then started a press campaign. Rawnsley, *Edward Thring*, p. 37.

34. *Lancet* (30 October 1875).

35. Probably a generic pen-name, and maybe not a single individual. I am indebted to Mr David Sharp, retired Deputy Editor of the *Lancet*, for his advice. This letter also appeared in other journals (e.g. the *Courier*); it was countered by 'A present Uppinghamian of seven and a half years' standing' in the same journal on 5 November 1875: copy in UA.

36. *Lancet* (30 October 1875). As the writer signed off by stating: 'I enclose my card', one assumes that the journal knew his identity.

37. Bell to Thring (n.d.), in UA.

38. *USM* (February 1916).

39. See Chapter 4.

40. *The Times*, 5 November 1875.

41. *Lancet* (6 November 1875).

42. *Liverpool Post*, 6 November 1875.

43. *Liverpool Daily News*, 6 November 1875.

44. An unnamed national daily paper, in Parkin (ed.), *Edward Thring*, vol. 2, p. 22.

45. Thomas William Jex-Blake (1832–1915), Principal of Cheltenham College, 1868–74, Headmaster of Rugby, 1874–87, later Dean of Wells, 1891–1911. By coincidence his sister, Sophia, was one of the pioneering figures in medicine as a career for women.

46. Thring to Jex-Blake, 6 November 1875, in Parkin (ed.), *Edward Thring*, vol. 2, p. 19.

47. *Lancet* (11 December 1875).

48. Tarbotton to Christian, December 1875, in UA. See also Hamlin, *A Science of Impurity*, pp. 152–211. 'The chemist Edward Frankland served as the quasi-official government analyst of the London water supply' (p. 153).

49. Frankland to Christian, 29 November 1875, in UA. Again, the analysis did not go beyond chemical issues.

50. *The Times*, 14 December 1875.

51. Memorandum from Barnard Smith, 6 January 1876, in UA.

52. Commentary by Parkin himself, in Parkin (ed.), *Edward Thring*, vol. 2, p. 21.

53. Maul to Thring, 25 December 1875, in UA.

54. Thring to Lambert, 3 January 1876, in NA MH12/9815.

55. Obituary article in *The Times*, 3 April 1900.

56. M. O. Tarbotton, 'Regulations etc for the Drainage of the School Houses' (1876), p. 4, in UA.

57. We do not know whether or not Thring had to put pressure on Hodgkinson to agree to be included in Tarbotton's survey, even though the Lower School was legally separate.

58. Uppingham School Trustees' Minute Books, 28 December 1875.

59. Thring to Christian, 10 January 1876, in UA.

60. See Chapter 1.

61. For example, near the Waggon and Horses public house.

62. Field, 'Report to the Sanitary Authority', p. 15.

63. Rawlinson, 'Uppingham', p. 2.

64. Uppingham School Trustees' Minute Books, 18 January 1876.

65. Thring to Sir Henry Thring, 19 January 1876, in Parkin (ed.), *Edward Thring*, vol. 2, p. 25. By this phrase he meant that the decision to allow the school to reassemble had been a close-run thing.

66. Ibid., pp. 26–7.
67. *Lancet* (21 January 1876); *Liverpool Daily News*, 22 January 1876.
68. *Lancet* (27 November 1875).
69. Thring's diary, 24 January 1876, in Parkin (ed.), *Edward Thring*, vol. 2, p. 28.
70. Thring's diary, 27 January 1876, in ibid., vol. 2, p. 28.
71. Haviland, 'Report on the Late Outbreak of Enteric Fever', p. 5.
72. Bell to LGB, 12 November, 7 and 21 December 1875, in NA MH12/9815.
73. James Alfred Wancklyn (1834–1906), analytical chemist, Professor of Chemistry at the London Institution, 1863–70, and public analyst for several towns and cities. See *Sanitary Record* (5 February 1876). See also Thring's diary entry for 7 February 1876, in Parkin (ed.), *Edward Thring*, vol. 2, p. 31.
74. Haviland, 'Report on the Late Outbreak of Enteric Fever', pp. 15–16. It is believed locally that it was known as the 'piss brook'.
75. Tate, 'West Deyne'.
76. See Rigby, unpublished manuscript, ch. 18, p. 8.
77. Report of the Commissioners: Schools Inquiry, vol. 16 (North Midland Division), p. 135.
78. Haviland, 'Report on the Late Outbreak of Enteric Fever', p. 27. Sing was in later life to become a headmaster himself as Warden of St Edward's School, Oxford.
79. Ibid., p. 43.
80. A leading member of the RSA. See Chapter 2.
81. Haviland, 'Report on the Late Outbreak of Enteric Fever', p. 9. See also UA: memorandum from Bell to Thring, 29 February 1876.
82. Smith, 'The Late Visitation of Typhoid Fever'.
83. Hodgkinson, 'Remarks on Mr Haviland's Report'.
84. W. Wales, 'A Letter to Rev RJ Hodgkinson Touching his Remarks on Mr Haviland's Report and the Action of the Sanitary Authority, 21 February 1876', in UA.
85. R. J. Hodgkinson, 'A Letter to the Rev Chancellor Wales in Reply to his Letter of February 21st, on the Action of the Uppingham Sanitary Authority' (25 February 1876), in UA.
86. W. Wales, 'A Rejoinder to the Second Letter of Rev RJ Hodgkinson on the Action of the Uppingham Sanitary Authority, 29 February 1876', in UA.
87. Registrar General's Annual Report, 1875: causes of death.
88. Dukes, *Health at School*. There had been a school doctor at Rugby since 1868. Dukes held the post from 1871 until 1908. See Introduction.
89. See Chapter 8.
90. Dukes, *Health at School*, pp. xv, 1. There is only a short list of sources cited in Dukes's bibliography, all written after 1876, and mostly about general school management. Dukes also writes that 'the time seems ripe for some adequate and simple guide to the application of the principles of hygiene to school life'. Searches of *BMJ* and *Lancet* in the years before 1875 reveal very little about school health management, other than factual reports of illness, and an article entitled 'Air in Public Schools' (*Lancet* (26 April 1873)) – but this was not about epidemics.

6 Spring 1876

1. Letter from Mrs Hodgkinson, in UA: copy given to the author by the late C. M. Hodgkinson, 1976.

2. Thring's diary, 28 January 1876, in Parkin (ed.), *Edward Thring*, vol. 2, p. 30.
3. Thring's diary, 29 January 1876, in ibid.
4. 29 January 1876, in Bell, 'Letterbook'.
5. *Lancet* (29 January 1876).
6. Thring's diary, 31 January 1876, in Parkin (ed.), *Edward Thring*, vol. 2, p. 31.
7. As Holden had done 25 years earlier, when he left Uppingham for Durham School.
8. *USR.*
9. Thring's diary, 18 January 1876, in Parkin (ed.), *Edward Thring*, vol. 2, p. 23.
10. *Stamford Mercury*, 4 February 1876.
11. *Stamford Mercury*, 25 February 1876.
12. *BMJ* (4 March 1876). See Chapter 5.
13. Handwritten copy of minute, in UA.
14. Thring's diary, 6 February 1876, in Parkin (ed.), *Edward Thring*, vol. 2, p. 31.
15. J. H. Skrine, *A Memory of Edward Thring* (London: Macmillan, 1889), p. 176.
16. Thring's diary, 20 February 1876, in Parkin (ed.), *Edward Thring*, vol. 2, p. 32.
17. Thring's diary, 22 February 1876, in ibid., p. 33. In Hoyland, *The Man Who Made a School*, p. 88, it is claimed that the navvies were so disgusted by what they found that they refused to complete the raising of the covers. This may, however, be based only on rumour.
18. Thring's diary, 23 February 1876, in Parkin (ed.), *Edward Thring*, vol. 2, p. 34.
19. 29 February 1876, in Bell, 'Letterbook'. See Chapter 5.
20. *Lancet* (26 February 1876).
21. Thring to Birley, 3 March 1876, in Parkin (ed.), *Edward Thring*, vol. 2, p. 35.
22. *Lancet* (4 March 1876).
23. 4 March 1876, in Bell, 'Letterbook'.
24. Bell to White, 5 and 7 March 1876, in ibid.
25. RSA resolution, 26 January 1876, in NA MH12/9816.
26. LGB to Thring, 17 January 1876, in ibid.
27. Rawlinson memorandum to LGB, 29 January 1876, in ibid.
28. W. H. Brown to LGB, 3 February 1876, in ibid.
29. Bell to LGB, 5 February 1876, in ibid.
30. Bell to LGB, 5 February 1876, in ibid.
31. Mullins to Haviland, 12 February 1876, in UA.
32. RSA to LGB, 9 March 1876, in NA MH12/9816
33. LGB internal memorandum, 18 February 1876, in ibid.
34. Joseph Rayner to LGB, 22 February 1876, in ibid.
35. Thring to LGB, 28 February 1876, in ibid.
36. Thring to LGB, 7 March 1876, in ibid.
37. Thring to LGB, 28 February and 7 March 1876, in ibid.
38. Haviland to LGB, 2 March 1876, in ibid.
39. LGB note on Brown's memorandum, 8 March 1876, in ibid.
40. Thring to LGB, 8 March 1876, in ibid.
41. Barnard Smith to LGB, 8 March 1876, in ibid.
42. LGB papers, 15 March 1876, in ibid.
43. RSA to LGB, 9 March 1876, in ibid.
44. See, for example, RSA to LGB, 6 January, and 4 and 24 February 1876, in ibid.
45. Thring to LGB, 7 March 1876, in ibid.

46. Skrine, *A Memory of Edward Thring*, p. 176. Skrine described it in romantic terms: 'Then someone spoke the right word: *"Don't you think we ought to flit?* The school can't stay here: let's take it somewhere else." There was a spark on tinder. "It's a big thing," thought we. "I'll do it!" said the chief.' Skrine was not a housemaster, but he appears to have been at the meeting, perhaps deputizing for someone else.

47. It was Campbell who was asked to oversee the school's interregnum between Thring's death in 1887 and the arrival of his successor.

48. Thring to Sir Henry Thring, 4 March 1876, in Parkin (ed.), *Edward Thring*, vol. 2, p. 37.

49. Thring's diary, 7 March 1876, in ibid., p. 39.

50. S. Haslam to Mr Copeman, 9 March 1876, in UA.

51. Thring to Johnson, 8 March 1876, in Parkin (ed.), *Edward Thring*, vol. 2, p. 41. Thring had sent one of the housemasters (Cobb) on a reconnaissance to investigate possible venues in several Welsh towns, including Llandrindod Wells, which he rejected as too small and bleak.

52. *Stamford Mercury*, 10 March 1876.

53. *BMJ* (26 February 1876).

54. *Lancet* (11 March 1876).

55. Thring to Birley, 5 March 1876, in Parkin (ed.), *Edward Thring*, vol. 2, p. 39.

56. Thring to Birley, 8 March 1876, in ibid., p. 40.

57. Skrine, *A Memory of Edward Thring*, p. 179.

58. Uppingham School Trustees' Minute Books, 11 March 1876. They were evidently unwilling to admit that, if one counted Hawke's death in June 1875, this was in fact the *third* outbreak.

59. Skrine, *A Memory of Edward Thring*, pp. 179–80.

60. *Stamford Mercury*, 10 March 1876; *Manchester Critic,* 10 March 1876; *The Times*, 11 March 1876.

61. Uppingham School Trustees' Minute Books, 11 March 1876.

62. Hoyland, *The Man Who Made a School*, p. 85, suggests that Thring threatened to resign if they prevented the school from moving; but again Hoyland gives no evidence.

63. Thring's diary, 11 March 1876, in Parkin (ed.), *Edward Thring*, vol. 2, p. 44.

64. *BMJ* and *Lancet*, both 11 March 1876.

65. Undated handwritten minute, in UA.

66. Thring's diary, 12 March 1876, in Parkin (ed.), *Edward Thring*, vol. 2, p. 44.

67. Thring to Jex-Blake, 13 March 1876, in ibid.

68. *The Times*, 14 March 1876.

69. Thring to LGB, 13 March 1876, in NA MH12/9816.

70. E. Thring, *Sermons Preached at Uppingham School*, 2 vols (Cambridge: Cambridge University Press, 1886), vol. 1, pp. 273–6.

71. Skrine, *A Memory of Edward Thring*, p. 177.

72. Ibid., p. 177.

73. Thring's diary, 12 March 1876, in Parkin (ed.), *Edward Thring*, vol. 2, p. 44.

74. Point made in conversation by Dr Elizabeth Hurren.

75. Bell to Brown, 18 March 1876 and Brown to Bell, 20 March 1876, in UA.

76. Undated notice to the Churchwardens, in UA.

77. A fishmonger, ironmonger, wine merchant, photographer, hairdresser, two tailors, two booksellers and three grocers. They included John Hawthorn and Charles White: the latter's involvement becomes significant in Chapter 7.

78. It would be different later in the year: see Chapter 7.

79. Undated note in UA, by letter on 23 and 24 March 1876, in Bell, 'Letterbook'.

80. *The Times*, 21–3 March 1876.

81. *Manchester Critic*, 24 March 1876.

82. Especially in view of the charges agreed with Mr Mytton, the owner of the Cambrian Hotel in Borth.

83. Uppingham School Trustees' Minute Books, 24 March 1876.

84. See Chapter 1. The boarding fees amounted to £21,000, the tuition fees to £9,000 and the charity income to £4,280.

85. The evidence on this issue is unclear. The Uppingham School Trustees' Minute Books, 24 March 1876, make no mention of what was to happen to these fees; given that Thring believed they were in the trustees' control rather than his (see Chapter 1), we have to assume that they were withheld. This would also seem consistent with Thring's later claim that he and the masters had incurred nearly £4,000 in lost fees: see Aftermath and Conclusion.

86. Thring's diary, 19 March 1876, in Parkin (ed.), *Edward Thring*, vol. 2, p. 46.

87. Unsigned document, 13 April 1876, in UA. It showed results from two boarding houses.

88. Mr C. M. Hodgkinson, a descendant, told the author in 1976 that he was sure it was the latter.

89. The setting-up work was led by Thring with military-style efficiency, with staff and pupils based in the Cambrian Hotel and two dozen small lodging-houses. A large wooden schoolroom was constructed for whole-school assemblies, and a specially chartered train brought furniture and equipment from Uppingham – even including the cricket roller. For details, see Skrine, *Uppingham by the Sea*, and *USM* (1876).

90. An undated account of events at this time, in UA, is signed 'Alice M. Bell' and may well have been written a long time later.

91. Hamlin, 'Mudding in Bumbledom'.

92. Uppingham Union Minute Book, 29 March 1876, ROLLR, DE1381/441.

93. Bell to M. H. Dobson, 29 April 1876, in Bell, 'Letterbook'.

94. Bell to Thring, 15 April 1876, in ibid. Childs had originally been recruited as sanitary officer, some months before the exodus to Borth was first mooted. See Chapter 4.

95. 30 March, 15 April and 3 June 1876, in Bell, 'Letterbook'. Bell wrote to Bagshawe about this watching brief, having feared for some time that Bagshawe lacked confidence in him. Bagshawe, however, replied accepting the idea quite readily.

96. Bell to Thring, 30 March 1876, in ibid.

97. It is unlikely that Thring fully appreciated the extremes of Atlantic weather in Borth, its remoteness or its high incidence of both scarlet fever and tuberculosis. It also had sanitation and water supply arrangements which were far from advanced.

7 Summer 1876

1. Thring to Christian, 26 March 1876, in UA.

2. *Manchester Critic*, 31 March 1876.

3. *Stamford Mercury*, 14 April 1876.

4. *USM* (May 1876).

5. 5 May 1876, in Bell, 'Letterbook'.

6. Surprisingly, despite his earlier support for the school, Foster was one of those targeted.

7.   Anonymous undated typed commentary, in UA.

8.   Candler to LGB, 30 March 1876, in NA MH12/9816.

9.   Ibid. The LGB agreed on the same day to an application from the RSA that Brown be given a bonus equivalent to his annual salary, in view of all his recent extra work.

10. He and his father both had close connections with the school: his shop was in High Street East.

11. See details of his attendance record in Appendix 1.

12. Bell to Jacob, 13 April 1876, in Bell, 'Letterbook'.

13. Bell to Brown, 6 May 1876, in ibid. Brown accused him of a vendetta. Bell denied this, but conceded the large extra workload resulting from the Public Health Act of 1875.

14. Bell to Thring, 27 April 1876, in ibid.

15. Bell to Beevor, 30 March 1876, in ibid.

16. Field to LGB, 4 April 1876, in NA MH12/9816.

17. Matthews, 'The New Water Supply', p. 26. It had bought a field west of the Kettering Road in March 1876, where it proposed to construct a well and water tower, from which mains would be laid into Uppingham. Separated from the town by two valleys, no pollution from drains or cesspits could have affected it.

18. Bell to Birley, 5 June 1876, in Bell, 'Letterbook'.

19. Undated memorandum from Haviland, in UA.

20. Bell later disputed these figures, having obtained others from meteorologist Mullins.

21. Bell to Beevor, 13 April 1876, and Bell to Revd J. C. Peake, 14 April 1876, in Bell, 'Letterbook'.

22. Bell to Beevor, 4 May 1876, in ibid.

23. Bell to Mullins, 5 May 1876, in ibid.

24. LGB internal memorandum, 7 May 1876, in NA MH12/9816.

25. Rawlinson memorandum, 8 May 1876, in ibid. It seems likely that he and John Lambert saw the situation very differently. See Chapter 2.

26. Bell to Jacob, 5 May 1876, in Bell, 'Letterbook'.

27. Bell to Thring, 16 May 1876, in ibid.

28. *Stamford Mercury*, 7 July 1876; copy of parliamentary bill, in UA.

29. The others were Benjamin Hopkins, a town draper, and William Garner Hart, a grocer. Hopkins was the only large-scale employer in the town, and also a leading dissenter. Hart was a grocer, dependent on the custom of the school.

30. LGB papers, 19 April 1876, in NA MH12/9816.

31. Drawings lodged at the Rutland Museum, 2005.

32. Field to Mullins, 5 May 1876, in UA.

33. Adderley seems to have changed his mind since his earlier opposition to the sewage farm. See Chapter 4.

34. Lord Gainsborough to LGB, 11 April 1876, in NA MH12/9816.

35. See Chapter 6.

36. RSA to LGB, 14 April 1876, in NA MH12/9816.

37. LGB papers, 17 April 1876, in ibid.

38. LGB to RSA, 25 April 1876, in ibid.

39. Withington to Charity Commissioners, 24 April 1876, in ibid.

40. See Chapter 6.

41. LGB to RSA, 25 April 1876, in NA MH12/9816.

42. RSA to LGB, 3 May 1816, in ibid.

43. *Stamford Mercury*, 12 May 1876. Noel was MP for Rutland; he was also a school trustee. See Chapter 2.

44. RSA resolution, 9 February 1876, in NA MH12/9816. It must, however, also be borne in mind that Uppingham was only one small part of the consortium of districts which employed him, and that it provided only a small proportion of his income.

45. Bell to Thring, 20 July 1876, in ibid.

46. RSA to LGB, 26 May 1876, in NA MH12/9816.

47. LGB to RSA, 7 June 1876, in ibid.

48. Bell to Thring, 2 August 1876, in Bell, 'Letterbook'.

49. Thring to Birley (following a letter from Bell), 7 July 1876, in Parkin (ed.), *Edward Thring*, vol. 2, p. 59.

50. Thring's diary, 7 July 1876, in ibid. Presumably this was based on information sent to him at Borth.

51. Bell to Thring, 6 July 1876, in Bell, 'Letterbook'.

52. Bell to Skrine, n.d., in ibid.

53. A detailed account of the school's time at Borth must await Thring's biographer. For details, see Skrine, *Uppingham by the Sea*; N. Richardson, 'Uppingham by the Sea: Typhoid and the Excursion to Borth, 1875–77', *Rutland Record*, 21 (2001), and N. Richardson, 'Uppingham's 1875–77 Typhoid Outbreak: a Re-Assessment of the Social Context', *Rutland Record*, 26 (2006). There is also much material in Parkin (ed.), *Edward Thring*, the *Cambrian News* and the *Aberystwyth Observer*. The National Library of Wales in Aberystwyth holds a number of secondary books describing life in Cardiganshire at that time. Events in Borth are comprehensively detailed in the *USM*. Skrine's lively if somewhat over-romanticized account of the school's time at Borth was published in 1878, soon after the school returned to Uppingham. In it he paints a faithful picture of a great adventure – helped by the generosity of a local landowner, Sir Pryse Pryse, who lent the school a games field, and performed many other acts of kindness. Thring felt that the outdoor life at Borth offered great educational experiences – and an opportunity to evade the increasing demands for more emphasis on team sports which the Old Boys had recently been making. However, life at Borth must have been hard for housemasters and their young, uprooted families.

54. Bell to Revd Godfrey Thring, 11 July 1876, in Parkin (ed.), *Edward Thring*, vol. 2, p. 60.

55. *Aberystwyth Observer*, 6 May 1876.

56. *Stamford Mercury*, 28 April 1876.

57. Thring's diary, 26 May 1876, in Parkin (ed.), *Edward Thring*, vol. 2, p. 64. Parkin also records (same page) that individual contributions continued to come in; Thring's diary of 29 November 1876 shows his gratitude for the gift of another £100 from Skrine 'as a birthday present for the school's expenses'.

58. Thring to an unnamed relative, n.d., in UA.

59. Thring to Birley, 15 June 1876, in UA.

60. Thring's diary, 22 June 1876, in Parkin (ed.), *Edward Thring*, vol. 2, p. 57.

61. Thring's diary, 1 July 1876, in ibid., p. 58.

62. Thring's diary, 26 June 1876, in ibid., p. 59.

63. Letter from Thring to Revd Godfrey Thring, 11 July 1876, in ibid., pp. 59–60.

64. All along I had said it was running our heads into a rat-trap; ... but I also told [Skrine] that I knew he thought me headstrong and impulsive, but that the bold dashes of resistance, when I made them, were the most solemn and deliberate acts of my life; that these

subjects were on my mind night and day; and that I never did anything dangerous without having very carefully counted the cost, made up my mind to possible defeat, and the more dangerous the more deliberate, at all events, my action was'. Thring's diary, 22 June 1876, in ibid., p. 57.

65. Thring's diary, 22 June 1876, in ibid., p. 57.
66. 26 June 1876, in Bell, 'Letterbook'. They were in a house on the Leicester Road.
67. Bell to Thring, 28 June 1876, in ibid.
68. Thring to Birley, 7 July 1876, in Parkin (ed.), *Edward Thring*, vol. 2, p. 59.
69. Uppingham School Trustees' Minute Books, 14 July 1876.
70. Thring's diary, 14 July 1876, in Parkin (ed.), *Edward Thring*, vol. 2, pp. 60–1.
71. Thring's diary, 20 July 1876, in ibid., p. 61. 'So the fight is over and the first battle won. The long day's struggle ended and happy dreams at last come down on Uppingham by the Sea. Altogether my heart is so full of gratitude.'
72. Birley to Bell, 7 June 1876, in UA.
73. *Aberystwyth Observer*, 19 August 1876. It took place at the home of a Captain Delahoy.
74. *Aberystwyth Observer*, 23 August 1876.
75. *Cambrian News*, 28 July 1876.
76. He would no longer be in Borth when these demands eventually yielded results a few years later.
77. It is interesting that Hodgkinson, having remained in Uppingham for the summer term, now visited Aberystwyth – possibly he was curious to see Borth for himself.
78. Drawn from letters and letterheads dated July and August 1876, in UA.
79. Bell to RSA, 6 July 1876, in Bell, 'Letterbook'.
80. Haviland to LGB, 1 July 1876, in NA MH12/9816.
81. Bell to LGB, 1 July 1876, in ibid. The house was owned by a Mr Peach, who had backed Bell.
82. RSA to LGB, 1 July 1876, and LGB to RSA, 2 July 1876, in ibid.
83. Bell to LGB, 4 July 1876, and LGB memorandum, 4 July 1876, in ibid.
84. LGB internal memorandum, 19 July 1876, in ibid.
85. Bell to Jacob, 2 August 1876, in Bell, 'Letterbook'.
86. RSA minute, 13 July 1876, in UA.
87. Bell to Thring, 26 July 1876, in Bell, 'Letterbook'.
88. See Chapter 2.
89. Note on the LGB enquiry, 5 July 1876, in UA.
90. Rawlinson to LGB, 10 July 1876, in NA MH12/9816.
91. LGB to RSA, 10 July 1876, in ibid; also 17 and 21 July 1876.
92. Bell to Skrine, 2 August 1876, in Bell, 'Letterbook'.
93. Bell to Jacob, 7 August 1876, in ibid.
94. Bell to Thring, 7 August 1876, in ibid.
95. Bell to Jacob, 9 August 1876, in ibid.
96. Bell to Sir Henry Thring, 10 August 1876, in ibid.
97. Thring to Christian, 9 August 1876, in UA.
98. Ibid., 'I am sure if the chancellor [rector] could be made to feel that the school would be broken up by the measures, and not simply that I should be hunted down, he would think twice over them'.
99. Bell to Jacob, 13 August 1876, in Bell, 'Letterbook'.
100. Bell to Jacob, 3 August 1876, in ibid.
101. He had been appointed before Thring's arrival in 1853.

102. LGB to RSA, 28 August 1876, in NA MH12/9816.

103. LGB to RSA, 28 August 1876, in ibid. See Matthews, *By God's Grace*, pp. 68–70, 83, 85, for a description of Earle; he appears to have been a good foil to Thring, in so far as he was loyal, cautious and steady.

104. Earle to Wales, 14 August 1876, in UA.

105. Earle to Christian, 14 August 1876, and Christian to Earle, from Ilminster, 14 August 1876, in UA.

106. Telegram and letter, 11 August 1876, in UA.

107. *Stamford Mercury*, 18 August 1876, confirmed in *Hansard*.

108. Christian to Clode, 16 August 1876, in UA. Clode held this post 1876–80, and was created Companion of the Bath in 1880. Bell confirmed this request in a letter to Jacob, 18 August 1876.

109. Hodgkinson to Christian, 16 August 1876, written from Marine Terrace, Aberystwyth, and Thring to Christian, 17 August 1876, written from Grasmere, both in UA.

110. Thring to J. C. Guy, 27 July 1876, in UA.

111. Thring to Christian, 14 August 1876, in UA.

112. Birley to Christian, 15 August 1876, in UA.

113. Birley to (probably) Thring, 16 August 1876, in UA. The letter is addressed to 'My Dear Sir'.

114. Kelly's *Directory*. Upcher was not a guardian; Piercy was, but went to only one meeting between 1875 and 1877.

115. Mullins to Christian, 14 August 1876, in UA.

116. See Chapter 6.

117. Bell to Jacob, 16 August 1876, in Bell, 'Letterbook'.

118. *Stamford Mercury*, 25 August 1876.

119. See Chapter 4.

120. I am indebted to Peter Lane for this point.

121. Bell to Jacob, 13 August 1876, in Bell, 'Letterbook'.

122. Bell to Jacob, 14 August 1876, in ibid.

123. Bell to Jacob, 16 August 1876, in ibid.

124. Bell to Jacob, 18 and 19 August 1876, in ibid.

125. See Chapter 8 for evidence of the pressure they were under by the time of the school's return. LGB papers show that the mortgage deed was finally signed and sealed by the RSA on 4 October.

126. Bell to Cogan, 16 August 1876, in ibid.

127. Bell to Jacob, 23 August 1876, in ibid.

128. Thring to Christian, 17 August 1876, in UA.

129. Thring to Christian, 25 August 1876, in UA.

130. Wortley to Thring 17 August 1876, in UA. He also wondered whether some recent remarks he had made about a small degree of typhoid being inevitable in Uppingham had been taken as meaning that he was unconcerned; if so, he wished to apologize.

131. Earle to Christian, 19 August 1876, in UA.

132. See Chapter 6.

8 Autumn, Winter and Spring 1876–7

1. *The Times*, 28 August 1876.

2. Bell to Jacob 30 August 1876, in Bell, 'Letterbook'.

3. The Registrar General's report confirms this: there were only 7 recorded typhoid deaths in 1876 – similar to other years in the decade, apart from 1875. Deaths overall were in line with earlier years.
4. *Stamford Mercury*, 1 September 1876.
5. Thring was uncomfortably aware that the lodging houses were very basic. Anticipating the worst, storm precautions were taken – including walling up the porch of the hotel with planks. Pupils faced the new term with rather less enthusiasm than in the heady days of April. With shorter days and the weather closing in, it was essential to find new amusements and diversions. Even Skrine admitted that there was some 'bullying and mischief'.
6. Thring's letter of thanks to Withington, 19 September 1876, in *USM* (October 1876).
7. See Aftermath and Conclusion.
8. *Lancet* (25 November and 2 December 1876). The allegations were robustly answered in a letter from Dr Childs a week later.
9. *USM* (October 1876).
10. *USM* (September 1876).
11. Bell to Earle, 19 September 1876, in Bell, 'Letterbook'.
12. Bell to Jacob, 1 September 1876, in ibid. 'Paterfamilias' wrote again to *The Times* on 5 September, reprinted in the *Stamford Mercury* three days later.
13. Bell to LGB, 19 September 1876, in NA MH12/9816.
14. Bell to Jacob, 1 September 1876, in Bell, 'Letterbook'.
15. Bell to Earle, 19 September 1876, in ibid.
16. Bell to Jacob, 21 September 1876, in ibid.
17. *Stamford Mercury*, 22 September 1876. Askew was the owner of the White Hart at 15 High Street West, and also a farmer.
18. Bell to Thring, 6 October 1876, in Bell, 'Letterbook'.
19. Thring's diary, 1 November 1876, in Parkin (ed.), *Edward Thring*, vol. 2, p. 63.
20. Bell to Jacob, 29 November 1876, in Bell, 'Letterbook'.
21. Bell to Christian, 15 November 1876, in ibid.
22. Bell to Thring, 2 December 1876, in ibid.
23. Bell to Christian, 2 and 6 December 1876, in ibid.
24. Bell to Thring, 2 December 1876, in ibid.
25. Bell to Christian, 1 December 1876, in ibid.
26. Bell to Christian, 8 December 1876, in ibid.
27. We cannot be sure whether or not Bell had suspicions about Brown's financial probity at this stage. See Aftermath and Conclusion.
28. Bell to Thring, 2 December 1876, and Bell to Christian 6 December 1876, in Bell, 'Letterbook'.
29. Bell to Christian, 19 February 1877, in ibid.
30. Bell to Candler, 1 December 1876, in ibid.
31. Bell to Candler, 1 December 1876, in ibid.
32. Haviland to RSA, 12 December 1876, in NA MH12/9816. It was sent to LGB, 22 December 1876. A draft had apparently been written some weeks earlier; see Haviland to RSA, n.d. [probably 8 November 1876], in ibid.
33. In some ways he personifies the modern phenomenon of a strong personality in a community who is determined to dictate local policy-making on a single issue and cannot let it go. Point made in conversation by Dr Elizabeth Hurren.
34. Bell to Candler, 14 December 1876, in Bell, 'Letterbook'.

35. *Stamford Mercury*, 22 December 1876.
36. *Lancet* (23 December 1876).
37. Thring to Guy, 27 September 1876, in UA.
38. Receipt, 30 September 1876, in UA.
39. Thring to Guy, 30 September 1876, in UA.
40. Thring to Guy, 20 October 1876, in UA.
41. *The Times*, 12 September, and 2 October 1876.
42. Bell to Thring, 6 October 1876, in Bell, 'Letterbook'.
43. Bell to Jacob, 29 November 1876, in ibid. A period of three weeks extra was granted.
44. Thring to LGB, 5 December 1876, and LGB memorandum attached, in NA MH12/9816/5.
45. Letter to his daughter, n.d. [probably December 1876], in UA.
46. Professor Henry Wentworth Dyke Acland (1815–1900), Honorary Physician to the Prince of Wales, Regius Professor of Medicine, Oxford University, 1857–94; Member of the Medical Council, 1854–74, President, 1874–87; Member of the Sanitary Commission, 1870–2; Companion of the Bath, 1883, Knight, 1884, Baronet 1890.
47. Acland to Thring, 20 December 1876, in UA.
48. Report from Acland, 20 December 1876, in UA.
49. Uppingham School Trustees' Minute Books, 22 December 1876.
50. Thring to Christian, 20 December 1876, in UA.
51. Letter to parents from the Cambrian Hotel, Borth, 26 December 1876, in UA.
52. Skrine, *Uppingham by the Sea*, p. 86. It is a less than generous judgement, given all the problems which Thring faced.
53. Uppingham Union Minute Book, 27 December 1877, ROLLR, DE1381/441.
54. Thring's diary, 6 January 1877, in Parkin (ed.), *Edward Thring*, vol. 2, p. 64.
55. Thring to Bell, 1 January 1877, in UA.
56. Uppingham Union Minute Book, 3 January 1877.
57. *Stamford Mercury*, 5 January 1877. It recorded later (23 March 1877) that Mrs Barnard Smith gave a stained glass window to Glaston Church in her late husband's memory, and returned to live in London with 'a very elegant gilt drawing-room clock and a handsome electro-plated inkstand', presented by the grateful parishioners of Glaston.
58. *Stamford Mercury*, 26 January 1877.
59. Bell to Jacob, 8 January 1877, in Bell, 'Letterbook'.
60. Uppingham Union Minute Book, 24 January 1877.
61. See Chapter 5.
62. Bell to Jacob, 10 February 1877, in Bell, 'Letterbook'.
63. Ibid.
64. Minute of RSA special meeting, 24 January 1877, in UA.
65. LGB internal memorandum, 3 January 1877, in NA MH12/9817.
66. Bell to Evans-Freke, 19 February 1877, in Bell, 'Letterbook'.
67. Uppingham Union Minute Book, 21 February 1877.
68. Bell to Jacob, 21 February 1877, in Bell, 'Letterbook'.
69. Bell to Christian, 27 January 1877, in UA.
70. *Stamford Mercury*, 23 March 1877.
71. The boys returned on 19 January 1877 – during howling gales which lasted a month. Storms were particularly violent at the end of January and again in late February; the school led a huge clearing-up operation the next day, and helped to repair breaches in

the sea-wall. Borth residents were fulsome in their praise. Not all the masters' wives and children were still there; some had resettled back in Uppingham.

72. Bell to Mullins, 20 February 1877, in Bell, 'Letterbook'.
73. Thring to Revd. AH Boucher, 21 February 1877, in UA.
74. Thring to Bell, 19 April 1877, in UA.
75. Bell to Mullins, 8 March 1877, in Bell, 'Letterbook'. Pateman, a solicitor, worked in the practice of William Sheild, which handled many of the school's property purchases, and was a close neighbour of Bell. See Chapter 1. Information from Peter Lane.
76. *Stamford Mercury*, 6 April 1877.
77. *Cambrian News*, 10 April 1877.
78. Bell to Jacob, 10 and 13 February, and 21 March 1877, in Bell, 'Letterbook'.
79. Bell to Thring and Jacob, both 21 March 1877, in ibid.
80. Bell to Thring, 26 March 1877, in UA.
81. Bell to Jacob, 21 March 1877, in Bell, 'Letterbook'.
82. *Stamford Mercury*, 6 April 1877.
83. Uppingham School Trustees' Minute Books, 16 February 1877.
84. Based on the *Stamford Mercury* reports, and on the bankruptcy notices in the *London Gazette*. The financial misfortunes of Thomas Freer (timber merchant and carpenter) on 9 May 1877 might possibly be connected with the school's absence, but there is no evidence. It may possibly have caused the bankruptcy of William Wilford (bookseller and stationer) on 14 November, but most, if not all, of the school's stationery needs would have been met by its staunch ally, John Hawthorn.
85. *Stamford Mercury*, 9 March 1877.
86. Skrine, *Uppingham by the Sea*, p. 107.
87. Thring's diary, 13 April 1877, in Parkin (ed.), *Edward Thring*, vol. 2, p. 69.
88. Thring's diary, 25 April 1877, in ibid.
89. Author and date unknown, in ibid., pp. 70–1.
90. Thring's diary, 3 May 1877, in ibid., p. 70.
91. Thring's diary, 7 May 1877, in ibid.
92. *Stamford Mercury*, 11 May 1877.
93. Bell to Jacob, 21 March 1877, in Bell, 'Letterbook'.
94. *Stamford Mercury*, 4 May 1877. See also Wales to Bell, 26 April 1877, in UA.
95. *USM* (Summer 1877).
96. *Stamford Mercury*, 22 June, 6 July and 9 November 1877.
97. Skrine, *A Memory of Edward Thring*, p. 187.
98. Bell to William Banks, 20 August 1877, in Bell, 'Letterbook'; Traylen, *Uppingham*, p. 24.
99. Bell to Bagshawe, 23 November 1877, in Bell, 'Letterbook'.
100. *Stamford Mercury*, 15 February 1877.
101. See Aftermath and Conclusion.
102. Candler to Bell, 11 September 1878, in UA. Candler was a director of the water company by then.
103. Undated document, in UA.
104. Ibid., p. 24.
105. Matthews, 'The New Water Supply', p. 26.
106. Traylen, *Uppingham*, p. 25. In one sense the reasons for the RSA's doubts and obstructions had been proved to be justifiable; water provision was then indeed a risky and inexact science.

107. Under the school's present playing field adjoining the Leicester Road. A water diviner helped with this discovery in 1892. See Matthews, 'The New Water Supply', p. 26.

108. Traylen, *Uppingham*, p. 25.

109. Graham, *Forty Years*, p. 7.

110. Traylen, *Uppingham*, p. 26. Just before the end of the century the company bought a new site to the right of the Gretton Road. It existed until 1956, when its directors finally handed it over to the Uppingham Rural District Council, the body which was in many ways the successor to the RSA.

111. Dukes, *Health at School*, pp. 270–1.

112. Bell to Thring, 7 March 1878, and Bell to Jacob, 7 and 23 December 1878, and 14 January 1879, in Bell, 'Letterbook'. Judging by the comments when he left Uppingham, Childs was a popular figure. See Aftermath and Conclusion.

113. Honey, *Tom Brown's Universe*, pp. 166–7; Honey's quote is from J. H. Simpson, *Schoolmaster's Harvest* (London: Faber and Faber, 1954), p. 124.

Aftermath and Conclusion

1. All tables drawn from 1871 census, and from Accounts and Papers: Local Government Taxation – Abstract of Sums Raised and Expended by Rural Sanitary Authorities, and Parliamentary Local Taxation Returns 1874–82. I am indebted to Dr John Davis for this line of enquiry.

2. 2nd Annual Report of the Public Works Loan Board, 20 June 1877 and subsequent years, in NA MH19/190.

3. See Chapter 3. Melton Mowbray was divided: it also had a USA in the mid-1880s.

4. The two larger towns had started accelerating their spending in the years before 1875: cf. Oakham, which saw a sudden leap in spending in 1879. Oakham had loan commitments which were nearly as big as Uppingham (*c.* £4,000) from 1878; by contrast, Market Harborough's loan was under £100; Melton Mowbray's was between £200 and £300 in this period.

5. Parkin (ed.), *Edward Thring*, vol. 2, p. 77. Thring suggested that the debt amounted to £3,000 in his diary on 12 February 1878 – but at this point he underestimated the true extent of his financial problems.

6. Rigby, unpublished manuscript, ch. 25 p. 3.

7. Uppingham School Trustees' Minute Books, 13 June 1877.

8. Ibid., 15 February 1878. See also ibid., 18 October 1877.

9. Ibid., 5 April 1878.

10. Ibid., 18 October 1878.

11. Thring, 'Petition of Edward Thring, Headmaster of Uppingham School, to the Charity Commissioners' (15 April 1878), in UA. The precise source of the additional £300 is not clear: probably raised by masters and other supporters.

12. Thring to the Charity Commissioners, 24 May 1878, in UA.

13. Charity Commissioners to Revd Sir J. H. Fludyer, 20 June 1878, in UA.

14. Undated note, in UA, and Uppingham School Trustees' Minute Books, 15 February 1878. The money had been owing from the date when it had been built in 1869.

15. Graham, *Forty Years*, pp. 161–2.

16. Matthews, *By God's Grace*, pp. 180–1.

17. Campbell, who is credited with the original idea to move the school, became temporary headmaster after Thring's death until a successor was appointed. John Skrine, having

failed to gain the post, left in 1888 to become warden of Glenalmond. In the early 1880s Hodgkinson left Uppingham altogether. Bagshawe became headmaster of the Lower School. It was absorbed into the school itself and became a senior boys' house in 1919, and was converted for use by girls in the early 1990s. In 1975, the greatly extended former sanatorium (Fairfield) had been put to the same purpose, having been little used in the recent years; the new sanatorium was a much smaller unit, reflecting the extent to which epidemics in schools had largely become a thing of the past. The workhouse on the Leicester Road closed in 1914, and briefly became an army hospital during the Great War. The site was sold to the school for demolition in 1923, and a new boarding house (Constables) was built there. See Traylen, *Uppingham*, p. 6.

18. Tate, 'West Deyne'.

19. 'It was most touching to hear my Upper Sixth telling me, sundry of them how happy they had been here especially in my division': Thring's diary 30 July 1880, in Parkin (ed.), *Edward Thring*, vol. 2, pp. 104–5. His reputation was secure; many visitors sought his advice. The school's tercentenary occurred in 1884. Thring became an author again. *Education and School* (London: Macmillan, 1864) had not been a commercial success, but *The Theory and Practice of Teaching* (Cambridge: Cambridge University Press, 1883) sold over 25,000 copies. One lecture, to Sunday School teachers, was held in the garden of Barnard Smith's rectory in Glaston (*Rutland, Oundle and Stamford Post* 2 July 1887).

20. Thring and others preached sermons which likened the year in Wales to the providential journeyings of the Israelites in the Old Testament.

21. Thring's diary, 23 September 1879, in Parkin (ed.), *Edward Thring*, vol. 2, p. 94. It is said that many Uppingham pupils who had been at Borth followed his example (*Cambrian News*, 17 November 1978). So did at least one housemaster (Sam Haslam), whose passion for fishing had been acquired there (*USM* (1908)). The Cambrian Hotel in Borth continued to dominate the coastline for a century until its demolition in the late 1970s. It housed a number of other educational institutions in times of war or other crisis – including physical education students from Chelsea College during the 1939–45 war. See also D. Stranack, *Schools at War: A Story of Education, Evacuation and Endurance in The Second World War* (Chichester: Phillimore, 2005). On its site there now stands a modern tourist office, with a small plaque next to the door recalling 'Uppingham: In grateful memory of 1876 and 1877'. The Uppingham Fields are still situated near the community hall and sports centre. Some of Thring's surviving Old Boys donated a fine east window to St Matthew's Church in 1925 which carries the inscription from the psalms: 'Thou shalt not be afraid, for the pestilence that walkest in darkness nor for the sickness that destroyeth in the noonday'. Uppingham itself would later play host to Kingswood School when the latter migrated from Bath for much of the 1939–45 war.

22. Thring to Christian, August 1881, in UA.

23. Confirmed by Cormac Rigby in conversation, 2005. There had been suggestions to Gladstone that he might be offered church preferment – possibly a canonry or deanery (Matthews, *By God's Grace*, p. 111), but nothing came of it.

24. *ODNB*, vol. 54, p. 683. Figures from the Calendar of the Grants of Probate, England and Wales. His estate was valued for probate at just over £5,000 in December 1887. Much of this probably resulted from the literary and other successes of his final years. On Saturday, 15 October 1887 he wrote in his diary: 'And now to bed, sermon finished, and a blessed feeling of Sunday coming' The final full-stop was never written. He was taken ill in the chapel on the following morning, and left the service. He died in School House six

days later. See also Rigby, unpublished manuscript, ch 28, p. 17. Thring inherited £6,728 from his grandfather, and after his father's death in 1875 there was a further £3,261. After Marie's death, his five children each received only £111.5.0. from his estate.

25. *The Times*, 28 October 1887.

26. *Stamford Mercury*, 4 November 1887.

27. Bell to Jacob, 21 June 1878, in Bell, 'Letterbook'. See also Chapter 7. Bagshawe had already moved his family to Childs's patient list. But Bell need not have worried; Childs fell in love with Thring's daughter, Margaret. Twelve years her senior, he was asked not to court her, but was discovered with her in the garden and dismissed. It seems likely that Childs's departure was greatly regretted by many in both school and town. See *USM* (1882).

28. *Medical Register* (1912). He also gained Child's post as the school's sanitary officer. See also E. W. Hornung, *Fathers of Men* (London: John Murray, 1912), pp. 90–1: in this novel Bell appears in fictional form as 'Old Hill', the school doctor who misdiagnoses a heart condition in the hero. Unlike Bell, he errs on the side of caution.

29. *USM* (1914 and 1915). The practice passed on his death to his son, William – who represented the fourth generation in the family to serve in this way. Bell's daughter Mary took up medicine herself and was briefly medical officer to Wycombe Abbey school before moving to hospital work in Norwich. See *The Medical Who's Who* (1935).

30. Preface to the first edition of Medical Officers of Schools Association, *The Handbook of School Health* (1884), in Cambridge University Library and on their website.

31. The 1881 census shows that he was still living in Northampton, St Giles district. No evidence has been found of him arousing similar controversies within other RSAs in his district. He appears in the 1891 Isle of Man census: Douglas Ward 6, and Manx Antiquarian Society Ledger, 1892–1908, in Manx National Heritage Library.

32. A. Haviland, 'Peel: Past, Present and Future – a Lecture in Aid of the Sailors' Shelter and Fishermen's Refuge', *Isle of Man Times*, 3 May 1884, 'The Manx Climate', *Lancet* (11 June 1887), 'The Necessity for Collecting and Arranging the Ascertained Facts relative to the Glaciation of the Isle of Man', *Isle of Man Times*, 6 February 1886, and 'The Isle of Man and the Glacial Period', *Isle of Man Times*, 16 January 1886, all in Manx National Heritage Library.

33. A. Haviland, 'The Essential Requisites of a Seaside Health Resort, and the Requirements of a Health Seeker, with the Physical Geography and Climate of the Isle of Man – a Lecture Delivered at the Masonic Hall, Douglas, 13 June 1883', and 'Consumption, the Social and Geographical Causes conducing to its Prevalence: Illustrated by a Coloured map – a Lecture Delivered at the Victoria Hotel, Douglas, 17 December 1883', in Manx National Heritage Library.

34. Revd T. Talbot, 'Mr Haviland's Lecture Reviewed', *Manx Sun*, 29 October 1883, and Talbot, 'Mr Haviland's Lecture and its Publication', *Manx Sun*, 7, 20 and 26 November, and 4 December 1883, in Manx National Heritage Library. See also *Isle of Man Times*, 16 June 1883. Talbot was a man who 'knew something about everything and everything about something ... frequently occupied in exposing errors made by writers ... he ruthlessly destroyed some of the most cherished Manx traditions ... in his zeal for historical accuracy, he was the perfect iconoclast'. See also obituary article on Talbot, *Manx Quarterly* (1902), p. 365, in Manx National Heritage Library.

35. A. Haviland, *Port St Mary as a Health Resort* (1891), in Manx National Heritage Library. He had written similarly about Brighton (1882) and Scarborough (1883). See also A.

Haviland, *Report on the Ballaquayle Building Estate, near Douglas Isle of Man* (Douglas, 1886), in Manx National Heritage Library.

36. A. Haviland, 'The Rising of the Sun at the Summer Solstice from Stonehenge', *Manx Astronomical Society* (1893), 'The Physical Geography of the Isle of Man: Inaugural Address to the Manx Geological Society, Established October 1888, Delivered 7 January 1889', *Isle of Man Times*, 26 January 1889, 'Inaugural Address on Medical Geography as an Aid to Clinical Medicine, also Phthsis and the Isle of Man – Delivered at the First Meeting of the Isle of Man Medical Society, 4 December 1896', *Lancet* (31 July 1897), and 'Mann or Man? – an Extract from the Report of the Proceedings of the Annual Meeting of the Isle of Man Natural History and Antiquarian Society, held at Douglas School of Art', *Isle of Man Times*, 29 March 1888, all held in Manx National Heritage Library.

37. The Manx Musuem in Douglas holds five examples of his work. He appears on the mainland in the 1901 census.

38. *Northampton Herald*, 13 June 1903; *BMJ* (27 June 1903).

39. Obituary articles in *Leamington Spa Courier* and *Northampton Herald*, 24 August 1889.

40. This resulted in his affairs going into liquidation in 1878, and a trial at the Oakham summer assizes in 1880 where 'the fountains of justice were poisoned ... [when] some indiscreet admirer had endeavoured to corrupt the body of jurors'. It could not be shown that Brown was involved in this, but the trial was moved to London, where he was subsequently acquitted in December. However, the Law Society struck him off in 1882 after his case was heard by two judges of the High Court (*Solicitors' Journal* (25 March 1882)). Wales's successor as Rector awarded the stewardship of the rectory manor to Frank Edward Hodgkinson, son of the former headmaster of the Lower School.

41. *USR* (eleventh issue, 1997), p. 31. Major. C. Bland was bursar from 1910.

42. Drawings lodged at the Rutland Museum, Oakham, 2005

43. Newsome, *A History of Wellington College*, pp. 215–19. It happened in 1891–2, and resulted in the college's temporary evacuation to Malvern.

44. R. Field, 'Byelaws and Regulations with Reference to House Drainage: Adopted by the Uppingham RSA and allowed by the LGB' (1878), in UA and Cambridge University Library.

45. Obituary article in *The Times*, 3 April 1900.

46. R. M. Jackson, *The Machinery of Local Government* (London: Macmillan, 1958), pp. 257 (footnote), 268. George Sclater Booth remained as head of the LGB until 1880. He played a major role in piloting the Public Health Act of 1879 through the House of Commons. During the 1880s his energies were directed towards local affairs in Hampshire, where he was a prominent magistrate and had business interests. A man of many enthusiasms, including hunting and shooting, art and music, he inherited Hoddington House and its estate in 1886, and accepted a peerage a year later, before becoming the first chairman of the new Hampshire county council in 1888. He died at Hoddington in 1894. *ODNB*, vol. 6, p. 616.

47. Szreter, *Health and Wealth*, p. 282.

48. See Chapter 4.

49. Webb and Webb, *Statutory Authorities for Special Purposes*, pp. 477–84. See also *ODNB*, vol. 57, pp. 810–15.

50. Garrard, *Leadership and Power in Victorian Towns*.

51. Highlighted in Chapter 7.

52. See Chapter 8.
53. Ibid., and Marshall, *Furness and the Industrial Revolution*.
54. Hennock, 'Finance and Politics', pp. 217–18.
55. See Chapter 8.
56. See Rogers, 'Prosperous – But Precarious'. This parallels what R. J. Morris has identified in the Paradise district of Oxford: see Morris, 'The Friars and Paradise'.
57. See A. Woods, *A Manufactured Plague? The History of Foot and Mouth Disease in Britain* (London: Earthscan, 2004), p. 140; Department for the Environment and Rural Affairs Report, 'Origin of the UK Foot and Mouth Disease Epidemic in 2001' (2002).
58. See Appendix 2.
59. See Robins, *The Story of Water Supply*, pp. 194–5.
60. See Millward and Sheard, 'The Urban Fiscal Problem'.
61. See Chapter 2.
62. See Bellamy, *Administering Central-Local Relations*.
63. See Chapter 2.
64. Wright and Hobhouse, *An Outline of Local Government*, pp. 28–31.
65. Haviland, 'Report on the Late Outbreak of Enteric Fever'. See p. 20 for his comments on ventilation.
66. Those of Field, Rawlinson and Haviland himself. Tarbotton's report was a purely technical engineering one.
67. Anon., 'Review of Geographical Distribution of Heart Disease, Cancer and Phthisis by Alfred Haviland'; Haviland, 'Report on the Late Outbreak of Enteric Fever', p. 20.
68. See Chapter 3.
69. See Chapter 5.
70. See Hamlin, *A Science of Impurity*, p. 205; Hassan, *A History of Water*, pp. 23–5, 48.
71. Digby, *The Evolution of British General Practice*, pp. 13, 65. See also Wilkinson, 'The Beginnings of Disease Control', pp. 23–8.
72. Digby, *Making a Medical Living*, pp. 300–1.
73. All this supports the views of Howkins, 'Rural Society and Community', p. 1501; and Bamford, 'Public School Town in the Nineteenth Century', pp. 25–36.
74. Wohl, *Endangered Lives*, p. 167. There is plenty of evidence of the existence of Mingay's 'middle-class elite', see Mingay, *Rural Life in Victorian England*, p. 167; also of Howkins's 'lieutenant class', see Howkins, 'Rural Society and Community', p. 1501.
75. The 'shopocacy' mindset (see Hamlin, 'Mudding in Bumbledom', p. 57) versus the school as patron/client (see Bamford, *Rise of the Public Schools*, pp. 192–3).
76. M. D. W. Tozer, *Physical Education at Thring's Uppingham* (Uppingham: Uppingham School, 1976), p. 177. This was the start of the period leading up to the Great War when the cult of 'manliness' became very prevalent in schools such as Uppingham. It is seen particularly in the emphasis laid on team sports and in the development of school cadet forces.
77. Leinster-Mackay, *The Educational World of Edward Thring*, p. 12.
78. *Stamford Mercury*, 28 October 1887.
79. For example, in Parkin's commentary. See Parkin (ed.), *Edward Thring*, vol. 2, p. 3.
80. Thring, 'Borth Commemoration Sermon'.

NOTE ON SOURCES

All the original documents listed as being in the Uppingham School archive are indexed and housed in filing cabinets. Because most of this work has been carried out fifty miles from Uppingham, they have been photocopied; the photocopies will be filed chronologically, and will be housed in the archive. All the texts discussed below are referenced in full at the relevant place in the notes, and appear in the Works Cited list.

Thring and the School

The school archive is very extensive, and Thring is well-documented. He has long been seen as a visionary figure – but an autocrat. Several of his staff wrote books about him after his death, some of which border on hagiography. He was also a skilled self-publicist. The archive contains many of his letters, and two years of his diaries, together with the selection of entries for other years published after his death by Sir George Parkin, a friend who was commissioned to produce a two-volume selection of extracts of diary entries and letters by Thring's family, who also ordered that the full diaries be destroyed. This selection needs to be treated with some critical care.

Thring wrote the diaries to leave a record for his children, but they also served as a means for him to dissipate his nervous energy and frustration with people and circumstances which he thought were obstructing him in realizing his vision. They were also something of a safety valve, as he tried to keep recurring money worries from his wife and children, preferring to confide them to the pages which he wrote up each night throughout his life. Perhaps it was inevitable that Parkin's selection dwells more on the struggles than the successes. The cumulative effect was to present a picture of someone depressed and anxious, sometimes bitter and petulant – rather than someone who simply needed to give vent to his emotions at the end of a long day. Parkin's selection does, however, provide a graphic illustration of the battles he had to fight over a long period, and the frustration he felt about those who opposed him.

The Town and its RSA

The publications of the Uppingham Local History Studies Group (ULHSG) relating to the town in 1802 and 1851 confirm its essential continuity of personnel, and (along with the 1871 census and local trade directories) provide much of the detail about its local networks. The Record Office for Leicestershire, Leicester and Rutland (ROLLR) holds the minute books of the Uppingham Union and the national school, as well as suit rolls and other documents, but not the minute book of the sanitary sub-committee. This appears to have been lost – either through flooding many years ago in the parish church, or through the later transfer of documents (first to Oakham, and thence to Leicester) as a result of periodic local government reorganization affecting Rutland.

The work of Auriol Thompson and Philip Pattenden on the chairman of the Uppingham guardians, Revd Barnard Smith, has added an important personal dimension to the administrative material. A number of technical drawings compiled during the period of the sewer system and sewage farm construction (some made by Rogers Field) have recently been lodged with the Rutland museum in Oakham.

The Local Government Board

The three boxes of National Archives (NA) papers relating to Uppingham which cover this brief period are an essential antidote to Thring's diaries, because they give us in-depth evidence of the dilemmas facing both the guardians and the LGB. In the year from October 1875 alone there were more than 150 papers or references to business between the LGB and either school or town, of which nearly 40 occurred in the final three months of 1875. They cover a huge range of minutiae – from vaccination records to audit matters and the remuneration of the schoolmaster at the workhouse, as well as recording in detail the bitter exchanges over the epidemic. Several other boxes have helped to establish the context in which both the LGB and the Public Works Loan Board (PWLB) operated.

The Doctors

Dr Bell's 'Letterbook' (UA) gives a spicy flavour of the rivalries between himself and the other local doctors, and of his strong resentment of Dr Haviland, and of the leading figures in the Uppingham union. It is the essential source for understanding the details of the complex, if short-lived, stand-off between school and town in the summer of 1876.

Haviland was a prolific writer, and two important reports – on the combined districts which he oversaw (1874) and on Uppingham itself (1875) – yield much evidence about his uncompromising character. Both the *Lancet* and the *BMJ* provide details of appointment and salaries of MOHs in other parts of the country in this period. They and other sources suggest that he was hardly typical of rural MOHs, in intelligence and medical training, even if counterparts elsewhere may have shared his zeal and strength of purpose. He had a large, but territorially illogical, geographical area spanning parts of four counties to oversee (explained in Chapter 3).

It has long been known that Haviland was highly critical of the school in his report, which raised issues well beyond those connected with typhoid and which was couched in inflammatory terms. What has hitherto been less evident is that he courted controversy in many of his other writings, and that there is more than a suspicion that he selected facts to fit preordained conclusions – a view which at least one reviewer or his work at the time shares with a modern writer. On the other hand, many of his recommendations would later be taken up as best practice for school doctors across the country.

Borth

Much of the detail is drawn from John Skrine's *Uppingham by the Sea,* published by Macmillan in 1878. It has been supplemented by numerous newspaper entries in the *Cambrian News* and the *Aberystwyth Observer*; the National Library of Wales in Aberystwyth also holds a number of secondary books describing life in Cardiganshire at that time. Events and personalities in Borth are comprehensively detailed both in Parkin's documentary selection and in the *USM*; they show how both staff and pupils viewed the unusual events that they were experiencing.

J. H. Skrine's lively if somewhat over-romanticized account of the school's time at Borth was published in 1878 soon after the school returned to Uppingham. In it he paints a faithful picture of a great adventure in which, having brilliantly improvised the living and teaching arrangements, staff and pupils worked cheerfully and purposefully together. While this description has much truth in it, it glosses over the hardships and practical difficulties, and understates the disagreements and less optimistic moments, especially in the later stages.

WORKS CITED

Primary and Local Sources

Cambridge University Library

Medical Officers of Schools' Association, *The Handbook of School Health* (1884).

Milner, T., 'Letter to the Revd. W. Wales in Reply to that Gentleman's Sermon, by Thomas Milner M.A., Northampton, 1838'.

Wales, W., *A Lecture Delivered in National School Room, Northampton to the Members and Friends of the Society, for the Diffusion of Religious and Useful Knowledge, Explanatory of the Nature and Objects of that Institution* (Northampton, 1839).

—, 'The Minister's Duty towards Himself and His People: A Sermon Preached at St. Giles's Church, Northampton, at the Visitation of the Lord Bishop of Peterborough Wed. 6 August 1851'.

Manx National Heritage Library, Douglas, Isle of Man

Isle of Man Census Records 1891 and 1901.

Manx Antiquarian Society Ledger, 1892–1908.

Haviland, A., 'The Essential Requisites of a Seaside Health Resort, and the Requirements of a Health Seeker, with the Physical Geography and Climate of the Isle of Man – a Lecture Delivered at the Masonic Hall, Douglas, 13 June 1883'.

—, 'Consumption, the Social and Geographical Causes conducing to its Prevalence: Illustrated by a Coloured Map – a Lecture Delivered at the Victoria Hotel, Douglas, 17 December 1883'.

—, 'Peel: Past, Present and Future – a Lecture in Aid of the Sailors' Shelter and Fishermen's Refuge', *Isle of Man Times*, 3 May 1884.

—, 'The Isle of Man and the Glacial Period', *Isle of Man Times*, 16 January 1886.

—, 'The Necessity for Collecting and Arranging the Ascertained Facts Relative to the Glaciation of the Isle of Man', *Isle of Man Times*, 6 February 1886.

—, *Report on the Ballaquayle Building Estate, near Douglas Isle of Man* (Douglas, 1886).

—, 'The Manx Climate', *Lancet* (11 June 1887).

—, 'Mann or Man? – an Extract from the Report of the Proceedings of the Annual Meeting of the Isle of Man Natural History and Antiquarian Society, held at Douglas School of Art', *Isle of Man Times*, 29 March 1888.

—, 'The Physical Geography of the Isle of Man: Inaugural Address to the Manx Geological Society, Established October 1888, Delivered 7 January 1889', *Isle of Man Times*, 26 January 1889.

—, *Port St Mary as a Health Resort* (1891).

—, 'The Rising of the Sun at the Summer Solstice from Stonehenge', *Manx Astronomical Society* (1893).

—, 'Inaugural Address on Medical Geography as an Aid to Clinical Medicine, also Phthsis and the Isle of Man – Delivered at the First Meeting of the Isle of Man Medical Society, 4 December 1896', *Lancet* (31 July 1897).

Talbot, T., 'Mr Haviland's Lecture Reviewed', *Manx Sun*, 29 October 1883.

—, 'Mr Haviland's Lecture and its Publication', *The Manx Sun*, 7, 20 and 26 November, and 4 December 1883.

National Archives, Kew

Local Government Board Papers relating to the Uppingham Poor Law Union 1860–1882, MH12/9812–17.

Local Government Reports/General Papers 1875–7, MH19/87–8.

Correspondence and Papers of the Local Government Board 1875–7, MH25/26–8.

Public Works Loan Board Papers 1835–92, MH19/190.

Northampton Central Library

Haviland, A., 'Report on the Geographical Distribution of Fever within the Area of the Combined Sanitary Authorities in the Counties of Northampton, Leicester, Rutland and Bucks 11 July 1874'.

Northamptonshire Record Office

Episcopal Visitation: Uppingham, 1878, ML598.

Oakham Public Library

Jones, C., 'Geology in Rutland', *Rutland Natural History Society Annual Report 2000*, Rutland Library Service Local Studies ref. 508.42545 (Stamford, 2000).

Record Office of Leicestershire, Leicester and Rutland

Barker's *Leicestershire and Rutland Directory* (1875)

Burial Registers, DE4862/1.

Census Return for Uppingham (1871), RG10/3301–2.

Harrod's *Directory of Leicestershire and Rutland* (1870).

Hunter's *Directory* (1848).

Kelly's *Directory of Leicestershire and Rutland* (1876).

Land tax assessment, 1874–5.

Matkins *Almanac* (1924).

Return of Owners of Land, vol. 2: Rutland (HMSO, 1873).

Slater's *Directory of Leicestershire and Rutland* (1850).

Statement of Expenses in the Restoration and Enlargement of Uppingham Church, DE5430.

Uppingham National School Minute Book, DE1784/64.

Uppingham Parish Church Restoration Fund, DE1784/23.

Uppingham Union Minute Book, DE1381/441.

Vestry Minute Book 1869, DE1784/24.

Wright, C. N., *Commercial and General Directory of Leicestershire and Rutland* (1880).

Uppingham School Archives (UA)

Manor of Preston and Uppingham Court Rolls, vol. 4.

Rectory Manor of Uppingham Court Rolls, vols 6 and 8.

Uppingham School Trustees' Minute Books.

Uppingham School Magazine (1853–87), bound volumes.

Uppingham School Roll (sixth issue, 1932; eleventh issue, 1997).

Bell, T., 'Letterbook' (1876–1904).

Field, R., 'Report to the Sanitary Authority' (6 January 1876).

—, 'Byelaws and Regulations with Reference to House Drainage: Adopted by the Uppingham RSA and Allowed by the LGB' (1878).

Haslam, Mrs S. L. E., Diary 1871–2.

Haviland, A., 'Report on the Late Outbreak of Enteric Fever in Archdeacon Johnson's School, Uppingham, Rutland: June–November 1875' (1876).

Hodgkinson, R. J., 'Remarks on Mr Haviland's Report on the Visitation of Fever in The School and Town of Uppingham' (10 February 1876).

—, 'A Letter to the Rev Chancellor Wales in Reply to his Letter of February 21st, on the Action of the Uppingham Sanitary Authority' (25 February 1876).

Jennings, J. C., 'Analysis of Uppingham Church Registers (n.d.).

Pidcock, J. H., 'A Report Presented to the Uppingham Sewer Authority Committee, on the Drainage, Water Supply etc, of the Town, by J. H. Pidcock Esq., Engineer' (May 1871).

Rawlinson, R., 'Uppingham: Town and School Reports' (1876).

Rigby, C., Unpublished manuscript of a book on Edward Thring (n.d.).

Rome, R. C., 'Uppingham: The Story of a School 1584–1948' (undated manuscript).

Smith, B., 'The Late Visitation of Typhoid Fever in the School and Town of Uppingham: A Statement of the Action of the Sanitary Committee by the Chairman 19 January 1876'.

Tarbotton, M. O., 'Regulations etc for the Drainage of the School Houses' (1876).

Tate, D., 'West Deyne: A Short History' (undated manuscript).

Thring, E., 'Borth Commemoration Sermon' (1880).

—, Diary 1859–62.

—, 'Petition of Edward Thring, Headmaster of Uppingham School, to the Charity Commissioners' (15 April 1878).

Wales, W., 'A Letter to Rev RJ Hodgkinson Touching his Remarks on Mr Haviland's Report and the Action of the Sanitary Authority, 21 February 1876'.

—, 'A Rejoinder to the Second Letter of Rev RJ Hodgkinson on the Action of the Uppingham Sanitary Authority, 29 February 1876'.

Government Sources

Accounts and Papers: Local Government Taxation – Abstract of Sums Raised and Expended by Rural Sanitary Authorities, and Parliamentary Local Taxation Returns, LXII–LXVII (1873–84).

Annual Reports of the Charity Commissioners, XX 13, 19 (1875–6).

Calendar of the Grants of Probate, England and Wales.

Department for the Environment and Rural Affairs Report, 'Origin of the UK Foot and Mouth Disease Epidemic in 2001' (2002).

Medical Research Council, *Epidemics in Schools: An Analysis of the Data Collected during the First Five Years of a Statistical Inquiry by the School Epidemics Committee* (London: HMSO, 1938).

Reports of the MOH of the Privy Council and Local Government Board.

Registrar General's Annual Reports.

Registrar General's Quarterly Bulletins.

Report of the Commissioners: Schools Inquiry, vol. 16 (North Midland Division).

Contemporary Journals

British Medical Journal.

Hansard.

Lancet.

Medical Officer.

Medical Register.

Private Schoolmaster.

Sanitary Record.

Servants' Magazine.

Solicitors' Journal.

Newspapers

Aberystwyth Observer.

Bridgwater Mercury.

Cambrian News.

Isle of Man Times.

Leamington Spa Chronicle.

Liverpool Daily News.

Liverpool Post.

London Gazette.

Manchester Critic.

Manx Quarterly.

Manx Sun.

Northampton District Chronicle.

Northampton Herald.

Rutland, Oundle and Stamford Post.

Stamford Mercury.

The Times.

Secondary Sources

Anderson, A. M., *The Antiseptic Treatment of Typhoid Fever* (Dundee, John Leng, 1892).

Anon., 'Review of Geographical Distribution of Heart Disease, Cancer and Phthisis by Alfred Haviland', *Athenaeum* (4 March 1876), p. 333.

Balls, F. E., 'The Origins of the Endowed Schools Act 1869' (PhD dissertation, Cambridge University, 1964).

Bamford, T. W., 'Public School Town in the Nineteenth Century', *British Journal of Education Studies*, 5 (1957), pp. 25–36.

—, *Rise of the Public Schools: A Study of Boys' Public Boarding Schools in England and Wales from 1837 to the Present Day* (London: Nelson, 1967).

Barrett, D. W., *Life and Work among the Navvies* (London: Wells Gardner, Darton and Co., 1880).

Barrett, F. A., 'Alfred Haviland's Nineteenth-Century Map Analysis of the Geographical Distribution of Disease in England and Wales', *Social Science and Medicine*, 46:6 (1998), pp. 767–81.

—, *Disease and Geography: The History of an Idea* (Toronto: York University Press, 2000).

Bartrip, P. W. J., *Mirror of Medicine: A History of the British Medical Journal* (Oxford: Clarendon Press, 1990).

Barty-King, H., *Water: The Book: An Illustrated History of Water Supply and Wastewater in the United Kingdom* (London: Quiller, 1992).

Bateman, J., *Great Landowners of Britain and Ireland 1876* (London: Harrison and Sons, 1878).

Beckett, J. V., *The Aristocracy in England* (Oxford: Basil Blackwell, 1986).

Bell, D., *Leicestershire and Rutland Privies: A Nostalgic Trip down the Garden Path* (Newbury: Countryside Books, 2000).

Bellamy, C., *Administering Central-Local Relations, 1871–1919: The Local Government Board in its Fiscal and Cultural Context* (Manchester: Manchester University Press, 1988).

Bennett, P., *A Very Desolate Position* ([Fleetwood]: Rossall School, 1977).

Binnie, G. M., *Early Victorian Water Engineers* (London: Telford, 1981).

Bourne, J., *Understanding Leicestershire and Rutland Place Names* (Loughborough: Heart of Albion Press, 2003).

Brand, J. L., *Doctors and the State: The British Medical Profession and Government Action in Public Health, 1870–1912* (Baltimore, MD: Johns Hopkins Press, 1965).

Brown, C., *Northampton 1835–1985: Shire Town, New Town* (Chichester: Phillimore, 1990).

Browne, G. F., *Recollections of a Bishop* (London: Smith, Elder, 1915).

Budd, W., *Typhoid Fever: Its Nature, Mode of Spreading, and Prevention* (London: Longmans, Green, 1873).

Burgess, G. H. O., *The Curious World of Frank Buckland* (London: Baker, 1967).

Cambridge, N. A., 'The Life and Times of Dr Alfred Carpenter 1825–1892' (DMed dissertation, University of London, 2002).

Carnell, G., *The Bishops of Peterborough 1541–1991* (Much Wenlock: R. J. L. Smith, 1993).

Cartwright, F. F., *A Social History of Medicine* (London: Longman, 1977).

Castle, E. B., *Moral Education in Christian Times* (London: Allen and Unwin, 1958).

Clark, P. (ed.), *The Transformation of English Provincial Towns 1600–1800* (London: Hutchinson, 1984).

Collins, E. J. T. (ed.), *The Agrarian History of England and Wales, VII, 1850–1914*, 2 vols (Cambridge: Cambridge University Press, 2000).

Creighton, C., *A History of Epidemics in Britain*, 2 vols (Cambridge: Cambridge University Press, 1891–4).

Daunton, M. J., *House and Home in the Victorian City: Working-Class Housing 1850–1914* (London: Edward Arnold, 1983).

—, *Trusting Leviathan: The Politics of Taxation in Britain 1799–1914* (Cambridge: Cambridge University Press, 2001).

Dickie, M., 'Liberals, Radicals and Socialists in Northamptonshire before the Great War', *Northamptonshire Past and Present*, 7:1 (1983–4), pp. 51–4.

Digby, A., 'The Rural Poor', in Mingay (ed.), *The Victorian Countryside* vol. 2, pp. 591–602.

—, *Making a Medical Living: Doctors and Patients in the English Market for Medicine, 1720–1911* (Cambridge: Cambridge University Press, 1994).

—, *The Evolution of British General Practice 1850–1948* (Oxford, Oxford University Press, 1999).

—, 'The Local State', in Collins (ed.), *The Agrarian History of England and Wales*, vol. 2, pp. 1425–64.

Dukes, C., *Health at School considered in its Mental, Moral and Physical Aspects* (London: Rivington, 1887).

Englander, D., *Poverty and Poor Law Reform in Nineteenth Century Britain 1834–1914: From Chadwick to Booth* (London: Longman, 1998).

Ensor, R. C. K., *England 1870–1914* (Oxford: Clarendon Press, 1936).

Everitt, A. (ed.), *Perspectives in English Urban History* (London: Macmillan, 1973).

Eyler, J. M., *Sir Arthur Newsholme and State Medicine 1885–1935* (Cambridge: Cambridge University Press, 1997).

Falkus, M., 'The Development of Municipal Trading in the Nineteenth Century', *Business History*, 19:2 (1977), pp. 139–46.

Finer, S. E., *The Life and Times of Sir Edwin Chadwick 1800–1890* (London: Methuen, 1952).

Fletcher, S., *Feminists and Bureaucrats: A Study in the Development of Girls' Education in the Nineteenth Century* (Cambridge: Cambridge University Press, 1980).

Foster, J., *Alumni Oxonienses* (Oxford: Parker and Co., 1888).

Frazer, W. M., *Duncan of Liverpool* (London: Hamish Hamilton, 1947).

—, A *History of English Public Health 1834–1939* (London: Bailliere, Tindall and Cox, 1950).

Freeman, T. W., 'Nineteenth Century Medical Geographer', *Geographical Magazine*, 10 (1978), p. 90.

Frost, J. C., *Haviland Genealogy* (New York: Higginson and Co., 1914).

Garrard, J., *Leadership and Power in Victorian Towns 1830–1880* (Manchester: Manchester University Press, 1983).

Gathorne-Hardy, J., *Doctors: The Lives and Works of GPs* (London: Corgi, 1984).

Goschl, K., 'A Comparative Study of Public Health in Wakefield, Halifax and Doncaster 1865–1914' (PhD dissertation, Cambridge University, 1999).

Graham, J. P., *Forty Years of Uppingham* (London: Macmillan, 1932).

Halliday, S., *The Great Stink of London: Sir Joseph Bazalgette and the Cleansing of the Victorian Capital* (Stroud: Sutton Publishing, 2001).

Hamlin, C., 'Muddling in Bumbledom: On the Enormity of Large Sanitary Improvements in Four British Towns 1855–1885', *Victorian Studies*, 32 (1988), pp. 55–83.

—, 'Politics and Germ Theories in Victorian Britain: The Metropolitan Water Commissions of 1867–9 and 1892–3', in R. M. Macleod (ed), *Government and Expertise: Specialists, Administrators and Professionals 1860–1919* (Cambridge: Cambridge University Press, 1988), pp. 111–23.

—, *A Science of Impurity: Water Analysis in Nineteenth Century Britain* (Bristol: Hilger, 1990).

—, *Public Health and Social Justice in the Age of Chadwick: Britain 1800–1854* (Cambridge, Cambridge University Press, 1998).

—, 'Sanitary Policing and the Local State, 1873–1874: A Statistical Study of English and Welsh Towns', *Social History of Medicine*, 18:1 (2005), pp. 37–61.

Hardy, A., *The Epidemic Streets: Infectious Disease and the Rise of Preventive Medicine 1856– 1900* (Oxford: Clarendon Press, 1993).

—, *Health and Medicine in Britain since 1860* (Basingstoke, Palgrave, 2001).

Hart, E., *Waterborne Typhoid: A Historic Summary of Local Outbreaks in Great Britain and Ireland 1853–1893* (London: Smith, Elder, 1897).

Hassan, J. A., 'The Growth and Impact of the British Water Industry in the Nineteenth Century', *Economic History Review*, 38 (1985), pp. 521–47.

—, *A History of Water in Modern England and Wales* (Manchester: Manchester University Press, 1998).

Haviland, A., 'Abstract of Two Lectures on The Geographical Distribution of Typhoid Fever in England and Wales: A Lecture Delivered at St Thomas's Hospital, London' *British Medical Journal* (10 February 1872), pp. 148–9.

—, *Hurried to Death or, a Few Words of Advice on the Dangers of Hurry and Excitement, Especially Addressed to Railway Travellers* (London, 1868).

Haw, G., 'The Local Government Board', *Contemporary Review*, 94 (1908), pp. 54–63.

Hennock E. P., 'Finance and Politics in Urban Local Government in England 1835–1900', *Historical Journal*, 6:2 (1963), pp. 212–25.

—, *Fit and Proper Persons: Ideal and Reality in Nineteenth Century Government* (London: Edward Arnold, 1973).

Hodgkinson, R. G., *The Origins of the National Health Service: The Medical Services of the New Poor Law 1834–1871* (London: Wellcome Historical Medical Library, 1967).

Holderness, B.A., 'Agricultural Responses to the "Depression" of the Late Nineteenth Century in England and France: Towards a Comparative History', paper presented to IV Congresso di Storia dei Movimenti Contadini – L'Agricultura en Europa e la masita della 'questione agrarian' 1880–1914 (Rome, 1993).

Honey, J. R. de S., *Tom Brown's Universe: The Development of the Victorian Public School* (London: Millington, 1977).

Hope Simpson, J. B., *Rugby since Arnold: A History of Rugby School from 1842* (New York: St Martin's Press, 1967).

Hopewell, J., *Shire County Guide to Leicestershire and Rutland* (Princes Risborough: Shire, 1984).

Hoppen, K. T., *The Mid-Victorian Generation 1846–1886* (Oxford, Oxford University Press, 1998).

Horn, P., 'Country Children', in Mingay (ed.), *The Victorian Countryside*, vol. 2, pp. 521–30.

Hornung, E. W., *Fathers of Men* (London: John Murray, 1912).

Howe, G. M., *Man, Environment and Disease in Britain: A Medical Geography through the Ages* (Newton Abbott: David and Charles, 1972).

Howkins, A., *Reshaping Rural England: A Social History 1850–1925* (London: Routledge, 1991).

—, 'Types of Rural Community', in Collins (ed.), *The Agrarian History of England and Wales*, vol. 2, pp. 1297–353.

—, 'Rural Society and Community: Overview', in Collins (ed.), *The Agrarian History of England and Wales*, vol. 2, pp. 1501–14.

Hoyland, G., *The Man Who Made a School: Thring of Uppingham* (London: SCM Press, 1946).

Huckstep, R. L., *Typhoid Fever and other Salmonella Infections* (Edinburgh: Livingstone, 1972).

Hunt, E. H., *Regional Wage Variations in Britain, 1850–1914* (Oxford: Clarendon Press, 1973).

Hurren, E.T., '"The Bury-al Board": Poverty, Politics and Poor Relief in the Brixworth Union, Northamptonshire c 1870–1900' (PhD dissertation, Leicester University, 2000).

—, 'Poor Law versus Public Health: Diphtheria, Sanitary Reform, and the "Crusade" against Outdoor Relief 1870–1900', *Social History of Medicine*, 18:3 (2005), pp. 399–418.

Irvine, A. L., *Sixty Years at School* (Winchester: P. and G. Wells, 1958).

Jackson, R. M., *The Machinery of Local Government* (London: Macmillan, 1958).

Jarman, S. G., *A History of Bridgwater* (Bridgwater: Elliott Stock, 1889).

—, *The Bridgwater Infirmary: A Record of its Rise and Progress* (St Ives: Jarman and Gregory, 1890).

Jenkins, A., *Rutland: A Portrait in Old Picture Postcards* (Seaford: S. B. Publications, 1993).

Kerr, B., 'Country Professions', in Mingay, *The Victorian Countryside*, vol. 1, pp. 288–99.

Kiple, K. F. (ed.), *The Cambridge Historical Dictionary of Disease* (Cambridge: Cambridge University Press, 2003).

Lambert, R., 'The Local Government Act Office', *Victorian Studies*, 6:1 (1962), pp. 121–50.

—, *Sir John Simon 1816–1904 and English Social Administration* (London: Macgibbon and Kee, 1963).

Lawes, J., 'Voluntary Schools and Basic Education in Northampton 1800–1871', *Northamptonshire Past and Present*, 6:2 (1979–80), pp. 85–91.

Leinster-Mackay, D., *The Rise of the English Prep School* (London: Falmer Press, 1984).

—, *The Educational World of Edward Thring* (London: Falmer Press, 1987).

Lewis, S., *A Topographical Dictionary of England*, 4 vols (London: S. Lewis and Co., 1848).

Longden, H. I., *Northamptonshire and Rutland Clergy from 1500*, 16 vols in 6 (Northampton: Archer and Goodman, 1938–52).

Loudon, I., *Medical Care and the General Practitioner 1750–1850* (Oxford: Clarendon Press, 1986).

Luckin, W., 'The Final Catastrophe – Cholera in London, 1866', *Medical History*, 21 (1977), pp. 32–42.

MacDonagh, O., *Early Victorian Government 1830–1870* (London: Weidenfeld and Nicolson, 1977).

Macleod, R. M., *Treasury Control and Social Administration: A Study of Establishment Growth at the Local Government Board 1871–1905*, Occasional Papers on Social Administration, 23 (London: Bell, 1968).

— (ed.), *Government and Expertise: Specialists, Administrators and Professionals 1860–1919* (Cambridge: Cambridge University Press, 1988).

Maltbie M. R., 'The English Local Government Board', *Political Science Quarterly*, 13:2 (1903), pp. 232–58.

Marshall, J. D., *Furness and the Industrial Revolution: An Economic History of Furness (1711–1900) and the Town of Barrow (1757–1897) with an Epilogue* (Barrow-in-Furness: Barrow-in-Furness Library and Museum Committee, 1958).

Matthews, B., 'The New Water Supply', *Borth Centenary Magazine* (1977), pp. 25–6.

—, *The Book of Rutland* (Buckingham: Barracuda Books, 1978).

—, *By God's Grace: A History of Uppingham School* (Maidstone: Whitehall Press, 1984).

McCrum, M., *Thomas Arnold: Head Master – A Reassessment* (Oxford: Oxford University Press, 1989).

Mee, A., *Leicestershire and Rutland* (Rotherham: King's England Press, 1997).

Millward, R., and S. Sheard, 'The Urban Fiscal Problem, 1870–1914: Government Expenditure and Finance in England and Wales', *Economic History Review*, 48:3 (1995), pp. 501–35.

Millward, R., and R. Ward, 'From Private to Public Ownership of Gas Undertakings in England and Wales, 1851–1947: Chronology, Incidence and Causes', *Business History*, 35:3 (1993), pp. 1–21.

Mingay, G. E., *Rural Life in Victorian England* (London: Heinemann, 1977).

— (ed.), *The Victorian Countryside*, 2 vols (London: Routledge, 1981).

Moberly, G., *George Moberly's Journal 1849*, ed. C. A. E. Moberly (London: John Murray, 1916).

Morris, R. J., 'The Friars and Paradise: An Essay in the Building History of Oxford 1801–1861', *Oxoniensia*, 36 (1971), pp. 72–98.

—, 'The Middle Class and the Property Cycle during the Industrial Revolution', in T. C. Smout (ed.), *The Search for Wealth and Stability: Essays in Economic and Social History Presented to M. W. Flinn* (London: Macmillan, 1979), pp. 91–113.

—, 'The Middle Class and British Towns and Cities of the Industrial Revolution 1780 to 1870', in D. Fraser and A. Sutcliffe (eds), *The Pursuit of Urban History* (London: Edward Arnold, 1983), pp. 286–305.

Nenadic, S., The Small Family Firm in Victorian Britain', *Business History*, 35:4 (1993), pp. 86–114.

Newman, B., *Portrait of the Shires* (London: Hale, 1968).

Newsome, D., *A History of Wellington College 1859–1959* (London: John Murray, 1959).

—, *Godliness and Good Learning: Four Studies on a Victorian Ideal* (London: John Murray, 1961).

Newton, D., and M. Smith, *The Stamford Mercury: Three Centuries of Newspaper Publishing* (Stamford: Shaun Tyas, 1999).

Page, W. (ed.), *The Victoria History of the County of Rutland*, 2 vols (London: Constable, 1908–35).

Palmer, R., *Folklore of Leicestershire and Rutland* (Wymondham: Sycamore Press, 1985).

Parkin, G. R., (ed.), *Edward Thring, Headmaster of Uppingham School: Life, Diary and Letters*, 2 vols (London: Macmillan, 1898).

Pattenden, P., and Thomson, A., 'The Snuffing of Sanitary Smith: Fellow and Senior Bursar', *Peterhouse Annual Record 2002–3* (2005), pp. 43–56.

Paul, J. A., *3000 Strangers: Navvy Life on the Kettering to Manton Railway* (Kettering: Nostalgia Collection, 2003).

Peach, A., *A Brief Account of the Uppingham Congregational Church and of the Fifty Years' Ministry of Rev. John Green* (Bournemouth, 1914).

Pelling, M., *Cholera, Fever and English Medicine 1825–1865* (Oxford: Oxford University Press, 1978).

Percival, A., *The Origins of the Headmasters' Conference* (London: John Murray, 1969).

—, *Very Superior Men: Some Early Public School Headmasters and their Achievements* (London: C. Knight, 1973).

Porter, D., *Health, Civilisation and the State* (London: Routledge, 1999).

Priestley, J., 'Conservancy versus Carriage Systems for the Disposal of Excreta', *Public Health*, 7 (1894–5), pp. 280–5.

Pyatt, H. R., *Fifty Years of Fettes* (Edinburgh: Edinburgh University Press, 1931).

Rawnsley, H. D., *Edward Thring: Teacher and Poet* (London: T. Fisher Unwin, 1899).

Rawnsley, W. F., *Edward Thring: Maker of Uppingham School, Headmaster 1853–1887* (London: Kegan Paul, Trench, Trubner and Co., 1926).

[Rawnsley, W. F.], *Early Days at Uppingham under Edward Thring by an Old Boy* (London: Macmillan,1904).

Reay, B., *Microhistories: Demography, Society and Culture in Rural England, 1800–1930* (Cambridge: Cambridge University Press, 1996).

Redlich, J., and F. W. Hirst, *The History of Local Government in England* (London: Macmillan, 1970).

Reynolds, R., *Cleanliness and Godliness: or The Further Metamorphosis* (London: George Allen and Unwin, 1943).

Richardson, N., 'Uppingham by the Sea: Typhoid and the Excursion to Borth, 1875–77', *Rutland Record*, 21 (2001).

—, 'Uppingham's 1875–77 Typhoid Outbreak: a Re-Assessment of the Social Context', *Rutland Record*, 26 (2006).

Rigby, C., 'The Life and Influence of Edward Thring' (DPhil. dissertation, Oxford University, 1968).

Rivington, W., *The Medical Profession: Being the Essay to which was Awarded the First Carmichael Prize of £200 by the Council of the Royal College of Surgeons, Ireland 1879* (Dublin, Fannin and Co., 1879).

Robins, F. W., *The Story of Water Supply* (Oxford: Oxford University Press, 1946).

Rogers, A., *The Book of Stamford* (Buckingham: Barracuda, 1983).

—, *The Making of Uppingham as Illustrated in its Topography and Buildings* (Uppingham: ULHSG, 2003).

—, 'Prosperous – But Precarious: Mortgages in the Local Economy of Uppingham in the Eighteenth and Nineteenth Centuries', *Family and Community History*, 8:2 (November 2005), pp. 105–22.

Ross, W. A., 'The Local Government Board and After: Retrospect', *Public Administration*, 13 (1956), pp. 17–25.

Royle, E., 'Charles Bradlaugh, Free Thought and Northampton', *Northamptonshire Past and Present*, 6:3 (1979–80), pp. 141–50.

Royle, S. A., 'The Development of Small Towns in Britain', in M. Daunton (ed.), *The Cambridge Urban History of Britain, Vol. III: 1840–1950* (Cambridge: Cambridge University Press, 2000), pp. 151–84.

Rutland Local History and Record Society, 'Who was Who in Rutland', *Rutland Record*, 8 (1988).

Saville, J., *Rural Depopulation in England and Wales 1851–1951* (London: Routledge and Kegan Paul, 1957).

Serjeantson, R. J., *A History of the Church of All Saints, Northampton* (Northampton: W. Mark, 1901).

Simmons, J., *Life in Victorian Leicester* (Leicester: Leicester Museums, 1971).

Simms, T. H., *The Rise of a Midland Town, Rugby, 1800–1900* (Rugby: Borough of Rugby Library and Museum Committee, 1949).

Simon, B., and I. Bradley, *The Victorian Public School: Studies in the Development of an Educational Institution* (Dublin: Gill and Macmillan, 1975).

Simon, J., *English Sanitary Institutions* (London: Smith, Elder, 1897).

Simpson, J. H., *Schoolmaster's Harvest* (London: Faber and Faber, 1954).

Skrine, J. H., *Uppingham by the Sea* (London: Macmillan, 1878).

—, *A Memory of Edward Thring* (London: Macmillan, 1890).

Smellie, K. B., *A History of Local Government* (London: Allen and Unwin, 1946).

Smith, B., *Arithmetic and Algebra in their Principles and Application* (Cambridge: Macmillan, 1853).

—, *Observations on the Ecclesiastical Dilapidations Act 1871 with a View to its Amendment or more Efficient Operation* (Cambridge, 1872).

Smith, F. B., *The People's Health 1830–1910* (London: Weidenfeld and Nicolson, 1979).

Smith, M., *The Story of Stamford* (Stamford: Martin Smith, 1994).

Squibbs, P. J., *A Bridgwater Diary 1800–1967* (Bridgwater: Somerset County Library, 1968).

Stenton, M. (ed.), *Who's Who of British Members of Parliament* , 4 vols (Hassocks, Harverster Press, 1976–81).

Stern, W. M., 'Water Supply in Britain: The Development of a Public Service', *Royal Sanitary Institute Journal*, 74 (1954), pp. 999–1005.

Strachey, L., *Eminent Victorians* (London: Penguin Classics, 1986).

Stranack, D., *Schools at War: A Story of Education, Evacuation and Endurance in the Second World War* (Chichester: Phillimore, 2005).

Sutherland, G., *Studies in the Growth of Nineteenth-Century Government* (London: Routledge and Kegan Paul, 1972).

Szreter, S., *Fertility, Class and Gender in Britain 1860–1940* (Cambridge: Cambridge University Press, 1996).

—, *Health and Wealth: Studies in History and Policy* (Rochester, NY: University of Rochester Press, 2005).

Thompson, F. M. L., 'Landowners and the Rural Community', in Mingay (ed.), *The Victorian Countryside*, vol. 2, pp. 457–74.

—, 'An Anatomy of English Agriculture, 1870–1914', in B. A. Holderness and M. Turner (eds), *Land, Labour and Agriculture, 1700–1920* (London: Hambledon Press, 1991), pp. 211–40.

Thomson, A., 'A Study of Roles and Relationships in a Rutland Village in the Mid Victorian Period: Glaston c1860–90' (MA dissertation, Leicester University, 1999).

Thring, E., *Education and School* (London: Macmillan, 1864).

—, *The Theory and Practice of Teaching* (Cambridge: Cambridge University Press, 1883).

—, *Sermons Preached at Uppingham School*, 2 vols (Cambridge: Cambridge University Press, 1886).

Topley, W. W. C., and G. S. Wilson, *Principles of Bacteriology, Virology and Immunology* (London: Edward Arnold, 1975).

Tozer, M. D. W., *Physical Education at Thring's Uppingham* (Uppingham: Uppingham School, 1976).

—, 'Thring at Uppingham-by-the-Sea: The Lessons of the Borth Sermons', *History of Education Society Bulletin*, 36 (1985), pp. 39–44.

—, 'Education for True Life: A Review of Thring's Educational Aims and Methods', *History of Education Society Bulletin*, 39 (1987), pp. 24–31.

Traylen, T., *Oakham in Rutland* (Stamford: Spiegl Press, 1982).

—, *Turnpikes and Royal Mail of Rutland* (Stamford: Spiegl Press, 1982).

—, *Uppingham in Rutland* (Stamford: Spiegl Press, 1982).

Uppingham Local History Studies Group (ULHSG), *Canon Aldred's Historical Notes* (Uppingham, 1999).

—, *Uppingham in 1851: A Night in the Life of a Thriving Town* (Uppingham, 2001).

—, *Uppingham in 1802: A Year to Remember?* (Uppingham, 2002).

—, *Uppingham at War: Uppingham in Living Memory: Snapshots of Uppingham in the Twentieth Century* (Uppingham, 2005).

Venn, J., and Venn, J. A., (comps), *Alumni Cantabrigienses: A Biographical List of all Known Students, Graduates and Holders of Office at the University of Cambridge, from the Earliest Times to 1900* (Cambridge, Cambridge University Press, 1947).

Von Glehn (Creighton), L., *Life and Letters of Mandell Creighton – by his Wife*, 2 vols (London: Longmans, Green, 1904).

Walford, E., *The County Families of the United Kingdom* (London: R. Hardwicke., 1895).

Waller, P. J., *Town, City and Nation: England 1870–1914* (Oxford: Clarendon Press, 1983).

Watkins, D., 'The English Revolution in Social Medicine 1889–1911' (PhD dissertation, University of London, 1984).

Webb, S., and B. Webb, *English Poor Law History Part II: The Last Hundred Years*, 2 vols (London: Longmans, Green and Co., 1929).

—, *Statutory Authorities for Special Purposes* (London: Longmans, Green and Co., 1922).

Wesley, J., *On Dress: A Sermon on I Peter III, 3,4* (Boston, MA: Samuel Avery, 1811).

Wilkinson, A., 'The Beginnings of Disease Control in London' (DPhil dissertation, Oxford University, 1980).

Williams, N., and G. Mooney, 'Infant Mortality in an "Age of Great Cities": London and the English Provincial Cities Compared', *Continuity and Change*, 9:2 (1994), pp. 185–212.

Wohl, A. S., *Endangered Lives* (London: Methuen, 1983).

Woods, A., *A Manufactured Plague? The History of Foot and Mouth Disease in Britain* (London: Earthscan, 2004).

Wright, R. S., and H. Hobhouse, *An Outline of Local Government and Local Taxation in England and Wales* (London: Sweet and Maxwell,1884).

Young, R. F. K. J., 'Sanitary Administration under the Local Government Board 1871–8' (BLitt. dissertation, Oxford University, 1964).

INDEX

N.B.: *t* after an entry indicates a table.